Synthesis Lectures on Power Electronics

This series publishes short books on topics related to power electronics, ancillary components, packaging and integration, electric machines and their drive systems, as well as related subjects such as EMI and power quality. Each Lecture develops a particular topic with the requisite introductory material and progresses to more advanced subject matter such that a comprehensive body of knowledge is encompassed. Simulation and modeling techniques and examples are included where applicable.

Ayan Mallik · Saikat Dey

Switching Modulator Optimization in Isolated Power Converters

Ayan Mallik
The Polytechnic School
Arizona State University
Mesa, AZ, USA

Saikat Dey
The Polytechnic School
Arizona State University
Mesa, AZ, USA

ISSN 1931-9525   ISSN 1931-9533 (electronic)
Synthesis Lectures on Power Electronics
ISBN 978-3-031-81575-1   ISBN 978-3-031-81576-8 (eBook)
https://doi.org/10.1007/978-3-031-81576-8

This Springer imprint is published by the registered company Springer Nature Switzerland AG
The registered company address is: Gewerbestrasse 11, 6330 Cham, Switzerland

If disposing of this product, please recycle the paper.

Preface

This book focuses on both the theoretical foundations and practical implementation aspects of optimizing switching modulation techniques in isolated power electronic converters. As new clean energy technologies emerge, converter system architectures are continually evolving, necessitating a deep understanding of switching strategies for various power converters to enhance system performance—particularly in terms of efficiency and miniaturization—before they are launched in the market quickly. This book aims to introduce, derive, and explain a wide array of switching techniques for the optimal design of diverse power converters, including multiport, multi-directional DC–DC, and DC–AC converters, which are widely used in applications such as transportation electrification, grid integration of renewables and storage, data center energy routing, solid-state transformers, aerospace, and space exploration.

The book opens with a generalized modeling approach for complex multi-order converters, establishing the relationship between bridge currents and control variables to accurately model soft-switching constraints across a variety of power transfer impedance characteristics. Chapter 2 starts with an analysis based solely on resistive non-ideality, while Chap. 7 extends this by incorporating full-order non-idealities, including transformer capacitive parasitics, into the modeling methods. Chapters 3–6 delve into the theoretical foundations of pulse width modulation (PWM), pulse frequency modulation (PFM), and phase modulation (PM) in both resonant and non-resonant multiport power electronic converters. These PWM and PFM techniques are widely applied in unidirectional and bidirectional DC–DC and AC–DC converters for diverse applications such as electric vehicle charging, grid-renewable integration, and space power supplies. Converters that operate over a broad voltage gain range often encounter efficiency challenges under extreme conditions, prompting the exploration of optimal switching modulation strategies to improve overall system performance, particularly efficiency across varying loads and gain ranges. The book highlights higher-order modulation techniques, such as phase-frequency modulation and phase-duty-frequency modulation, which leverage all control variables as degrees of freedom to enhance converter efficiency by minimizing RMS current and expanding the soft-switching region. The book also covers the closed-loop integration of multi-order modulators with the PWM or PFM compensators, emphasizing their synthesis and formulation, the constraints of microcontroller implementation, and strategies for hardware realization, all as part of the optimization and development of advanced high-frequency power converters.

Mesa, AZ, USA   Prof. Ayan Mallik
October 2024

Acknowledgements

Many of the concepts and methods presented in this book have developed through discussions with ASU Power Electronics and Control Engineering (PEACE) lab's graduate students, alumni, interns, and project collaborators over the past few years.

We wish to acknowledge that this book would not have been possible without the support of the organizations that funded our research over the past years. We are especially grateful for the opportunity to collaborate with remarkable colleagues, whose support, technical discussions, feedback, and contributions have significantly enriched our research journeys that have been instrumental in the production of the material presented in this book.

We extend our gratitude to Dr. Arza Seidel for initiating the conversation about this opportunity and support in publishing this book. We are also grateful to the Springer Editorial team for their careful scrutiny of the material presented in this book.

Lastly, we are extremely thankful to our families and friends for their unwavering support, both during the writing process and throughout the numerous challenges that come with academic life.

Contents

Isolated Power Electronic Converters

1

1.1   Isolated PWM-Controlled Two-Port Converters

The rapid advancement in technology and the growing demand for efficient and reliable power management have driven the development and deployment of advanced isolated DC-DC converters. These converters play a crucial role in various emerging applications by providing effective solutions for power conversion and management. Among the different types of isolated DC-DC converters, the Dual Active Bridge (DAB), LLC, CLLC resonant converters, and Phase-Shifted Full-Bridge (PSFB) converters stand out for their unique advantages and wide range of applications.

DAB converters, shown in Fig. 1.1, due to their network symmetry on both sides of the isolation transformer, are known for their bidirectional power flow capabilities, which make them particularly suitable for energy storage systems and electric vehicle (EV) chargers. Their ability to handle high power densities and facilitate efficient power transfer between different voltage domains positions them as a valuable asset in renewable energy integration and grid energy storage solutions. Aside from dc-dc conversion, DABs are also utilized in isolated dc-ac or ac-dc grid-connected power conversion, such as PV to grid, battery-to-grid inverters, grid-connected front-end power factor correction (PFC) rectifiers, as illustrated in Fig. 1.2a–c.

Among the other types of traditional dc-dc converters, phase-shifted full bridge (PSFB) converters (shown in Fig. 1.3a–b) have garnered considerable attention for medium to high-power applications due to their straightforward design and zero-voltage switching (ZVS) capability [1]. Unlike resonant DC-DC topologies, the voltage gain of a PSFB converter is not influenced by the switching frequency but instead varies with the phase-shift angle between the duty cycles of diagonally opposed primary-side switches. Achieving high conversion efficiency requires maintaining ZVS across a broad range of load powers. A significant challenge in designing a PSFB converter is ensuring ZVS under light-load

© The Author(s), under exclusive license to Springer Nature Switzerland AG 2025
A. Mallik and S. Dey, *Switching Modulator Optimization in Isolated Power Converters*,
Synthesis Lectures on Power Electronics, https://doi.org/10.1007/978-3-031-81576-8_1

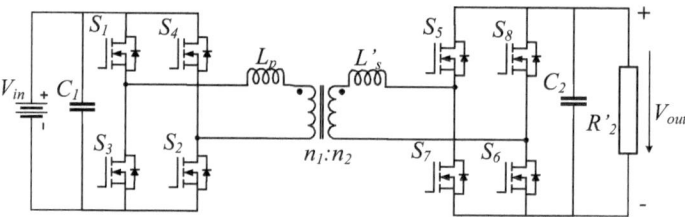

Fig. 1.1 DAB converter topology schematic

conditions, which is sensitive to the inductance value and switching frequency at lower power levels. Increasing the resonant inductance can facilitate ZVS at light loads, but this adjustment can lead to drawbacks such as duty cycle loss on the secondary side, additional dead-time conduction losses, and increased ringing across the secondary-side rectifier [2, 3]. Therefore, an actively controlled method where adjustment to any passive parameters or incorporation of any auxiliary circuit is not necessary would be of significant interest. It is also to be noted that PSFB converters are more of a popular choice when it comes to unidirectional power flow. That is mainly because a center-tapped configuration of PSFB, as shown in Fig. 1.3b, uses a lower number of semiconductors on the secondary side and could turn into a more cost-effective design than a DAB. One drawback of PSFB is the requirement of an output DC high-flux inductor which can adversely affect power density enhancement, as it adds bulk and weight to the system.

In the above-mentioned pulse width modulated (PWM) dc-dc and dc-ac converter topologies, the controllable switches are operated in a switch mode where they are required to turn on and turn off the entire phase current during each switching. In this switch mode operation, unless any advanced modulation strategies are adopted, the inherent trend is that the devices are subjected to high switching stresses and high switching power loss that increases linearly with the PWM switching frequency. Another significant drawback of the switch mode operation is the electromagnetic interference (EMI) produced due to large di/dt and dv/dt, especially with the use of wide bandgap semiconductors that have smaller junction capacitances and hence greater slew rates for switching transitions [4]. Therefore, at the present age with pressing needs for high power density and miniaturization, it is critical to explore high-performance modulation and control strategies that would enable soft-switching in the PWM-controlled converters which would allow one to design the converters for higher frequencies without severely compromising efficiencies, such that the passive components specifically magnetic and dielectric elements can be miniaturized.

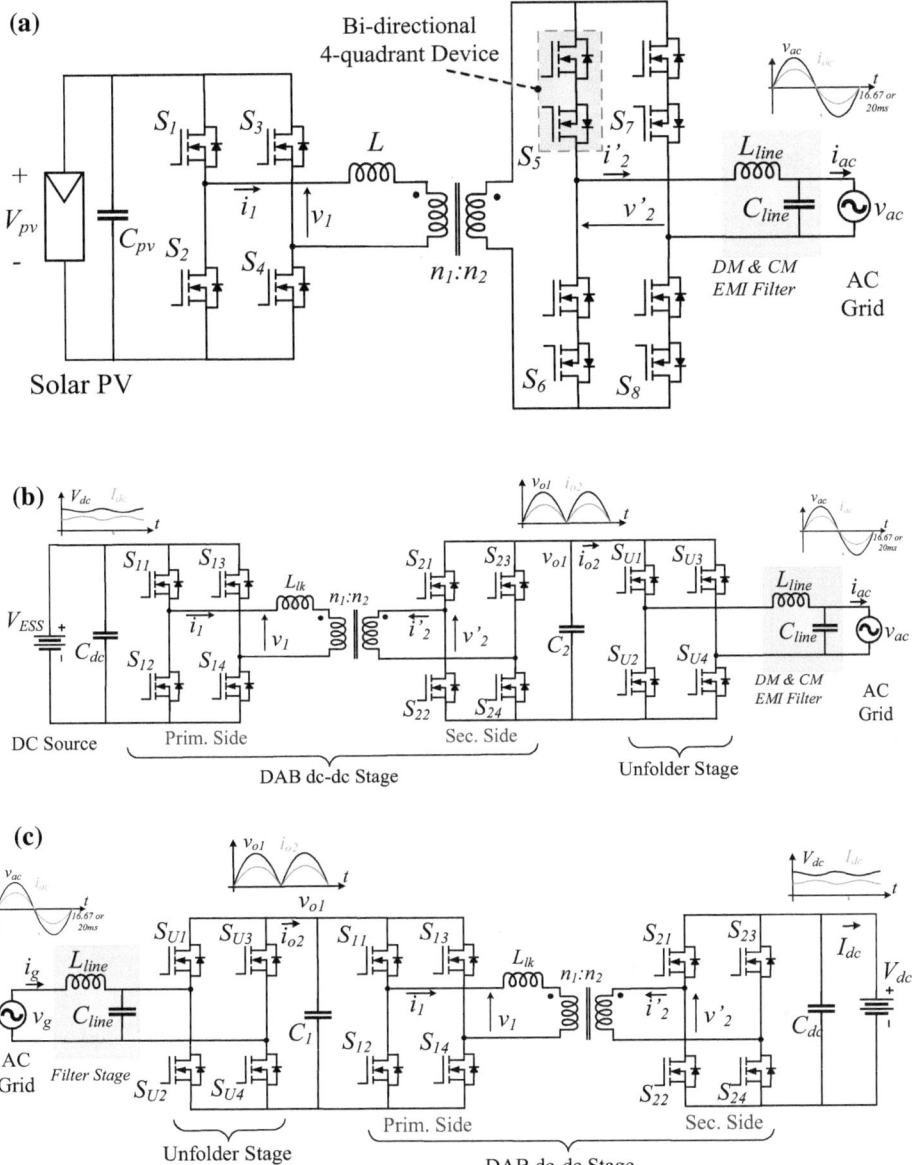

Fig. 1.2 a DAB dc-ac isolated converter interfacing PV to grid using bidirectional four-quadrant power devices. **b** DAB dc-ac isolated converter interfacing energy storage system to grid using line frequency unfolder structure on the AC side. **c** DAB ac-dc power factor correction converter topology with intermediate AC link

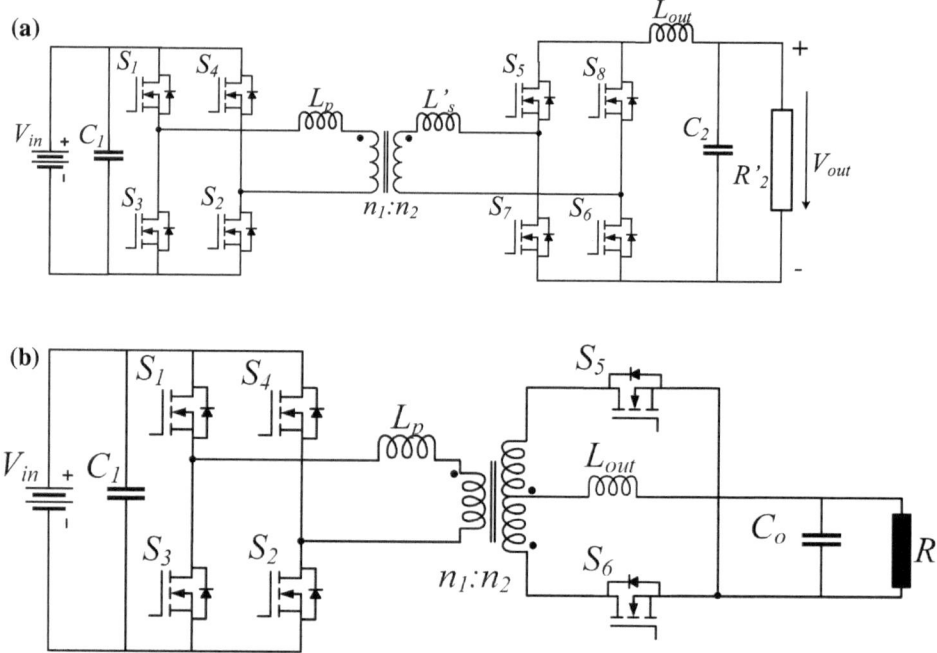

Fig. 1.3 **a** PSFB dc-dc topology with full-bridge secondary side configuration. **b** PSFB dc-dc topology with center-tapped secondary side configuration

1.2   Isolated PFM-Controlled Two-Port Converters

A parallel avenue for achieving high-density power conversion is to use resonant converters that have inherent features of soft-switching and hence naturally allow the switching frequencies to be considerably higher without much impact on the power losses and thermal management needs. LLC and CLLC are widely used as resonant converter candidates in EV charging, renewables integration, wireless power transfer and many other stationery and transportation applications. An LLC converter (as shown in Fig. 1.4) is typically controlled on its primary side through pulse frequency modulation (PFM), while a synchronous rectification (SR) control is performed on the secondary side that mimics diode-bridge behavior and achieves zero current turn-off and hence zero turn-off losses. The PFM ensures the converter to be operating in the inductive zone seen from the primary side which is favorable for achieving ZVS turn-on and hence zero turn-on losses. The only two switching loss types to be accounted for are primary turn-off and secondary CV^2 loss, which are still proportional with switching frequency. Therefore, this necessitates an optimization routine to explore if any multivariable control involving the bridge duty ratios, switching frequency, and phase shift can be established to minimize

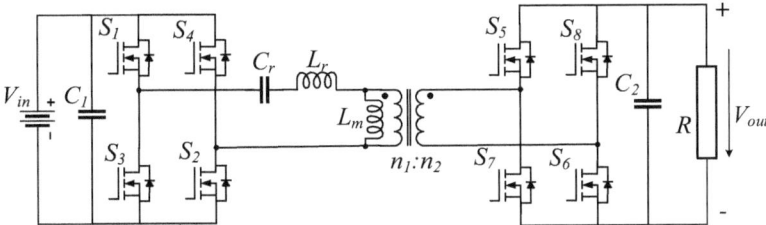

Fig. 1.4 Full-bridge LLC converter schematic

the net switching network losses. If so, it is further critical to understand the real-time implementation details in a particular converter use-case scenario.

It is important to note that LLC is mostly suitable for unidirectional power conversion because in the reverse power flow direction, its equivalent network turns out to be very similar to a series resonant converter (SRC) [5] which limits the normalized voltage gain below unity, hence limiting from being able to do any voltage boosting operation. Therefore, LLC's capability in wide-gain bidirectional power flow is restricted. In fact, CLLC converter, as shown in Fig. 1.5, is a suitable candidate as a replacement of LLC when it comes to bidirectional power transfer due to its symmetric tank structure on the two sides of the transformer. LLC is still preferred over CLLC in unidirectional power flow due to no requirement of secondary resonant capacitor bank, making it more reliable and slightly more power dense. In any power flow direction, the source side full-bridge in a CLLC achieves ZVS turn-on, while the load side full-bridge achieves SR when the switching frequency is very close to resonant frequency and is capable to achieve ZVS otherwise [6]. The full-bridge CLLC topology offers four degrees of freedom, which are the switching frequency, two bridge duty ratios, and the phase shift angle. It is critical to note that there is only one mandatory constraint related to power transfer or voltage regulation in a two-port CLLC, which means only one control variable can be engaged in that while the other three are freely available for performing multidimensional optimization in the converter. The sense of optimization becomes even stronger when the switching frequency deviates from the resonant point due to non-fundamental harmonics being more prominent in power transfer, especially in a wide-gain CLLC or LLC, primarily because the there are many feasible control variable sets that enable the same power conversion, but it is important to operate the converter at the most optimum set that would maximize the range of soft-switching and minimize the RMS currents. Exploring this would first require us to dive deep into the modeling techniques of resonant converters and perform required corrections in the existing modeling approaches with no approximations or assumptions in place.

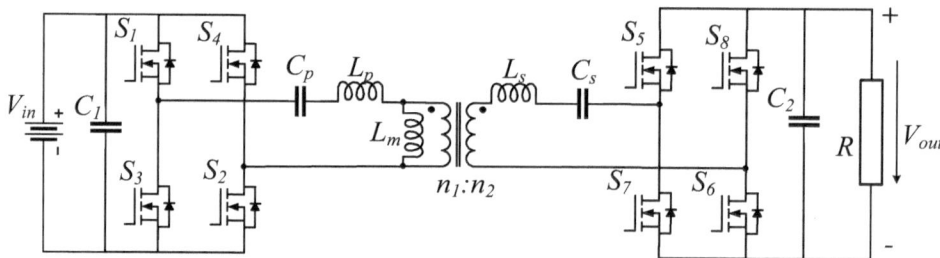

Fig. 1.5 Full-bridge CLLC converter schematic

1.3   Isolated PWM/PFM-Controlled Multi Active Bridge Converters

While the descriptions above mainly are targeted towards two-port power converters, recently there are emerging applications and needs of multi active bridge (MAB) converters as a promising solution for versatile energy management systems that connect multiple source and load ports together in a unified interface and enable multidirectional power flow with enhanced degree of modularity. An n-port MAB converter consists of n full-bridge or half-bridge modules that are magnetically coupled through an n-winding high-frequency (HF) transformer. The key benefits of using an MAB converter for versatile energy management include: (a) the ability to integrate multiple electrical sources and loads with minimal dc-dc conversion stages, thereby enhancing power density by reducing the number of components; (b) the adjustable transformer turns ratio, which supports various sources and loads with different voltage levels; (c) bidirectional power flow across all ports, enabling efficient and reliable power sharing; and (d) zero-voltage switching (ZVS) capability of the active bridges, which minimizes switching losses across a wide range of operations [7]. Due to these advantages, MAB converters are widely used in applications such as distributed renewable energy systems, energy router power conversion, solid-state transformers [8], electric vehicle (EV) on-board chargers [9], and more-electric aircraft power systems [10].

An example of a four-port power conversion system integrating various distributed assets such as solar PV string, energy storage system, DC distribution bus, and an AC grid is furnished in Fig. 1.6. Such a power conversion can also be facilitated by a combination of several two-port converters depending on the power flow modes and directions, as shown in Fig. 1.7, but that would increase the total number of semiconductors, transformers and other magnetic elements, and thermal management systems, therefore degrading the net power density and also increasing the bill-of-material (BOM) cost of the system. Furthermore, that would require multiple decentralized control systems and hence a reliable and robust communication link to be established between them. A modular

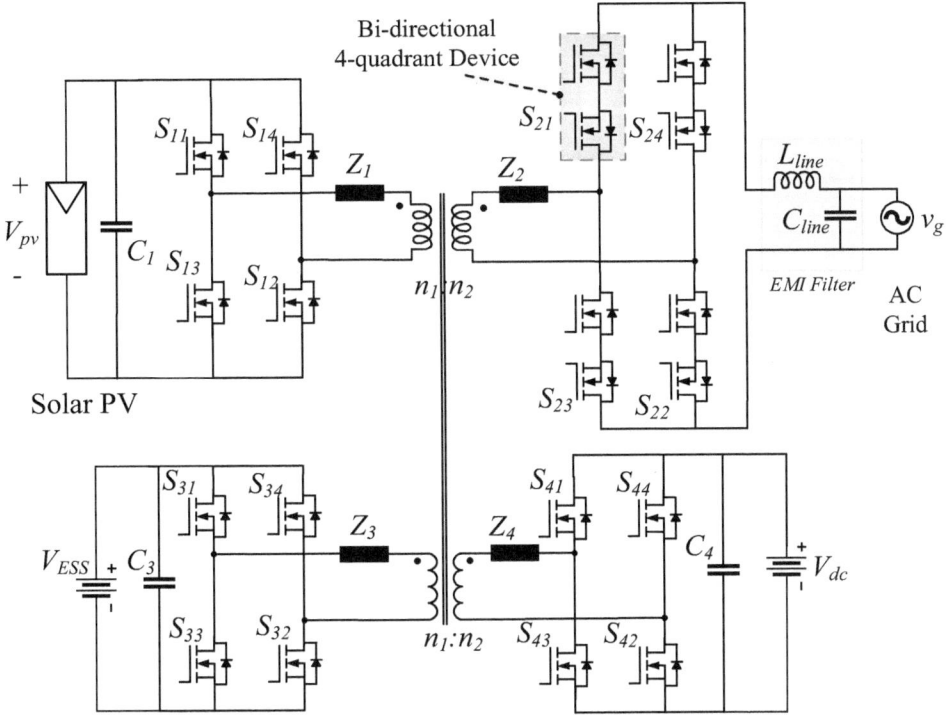

Fig. 1.6 Four-port integrated dc-ac-dc multidirectional power conversion interfacing PV, ESS, DC bus, and AC grid

and integrated system realized by MAB, on the other hand, can take advantage of a centralized control and communication system and would utilize one single multi-winding transformer. However, one of the major challenges behind the development of an MAB converter is to maintain its efficiency at several corner conditions both in terms of load power and voltage gains at different ports. While a design level optimization can be carried out in the development phase that can account for soft-switching realizability and loss minimization, that would only target one or several discrete operating points and not the entire spectrum of load and gain conditions. Therefore, in practice, during converter operation, the switching modulation strategy involving the available control variables must be adaptively optimized and executed in real time for converter performance enhancement. In the following chapters, we analyze and examine the correlation of different power loss functions with control variables and introduce and validate the optimization framework and implementation details in digital control platforms.

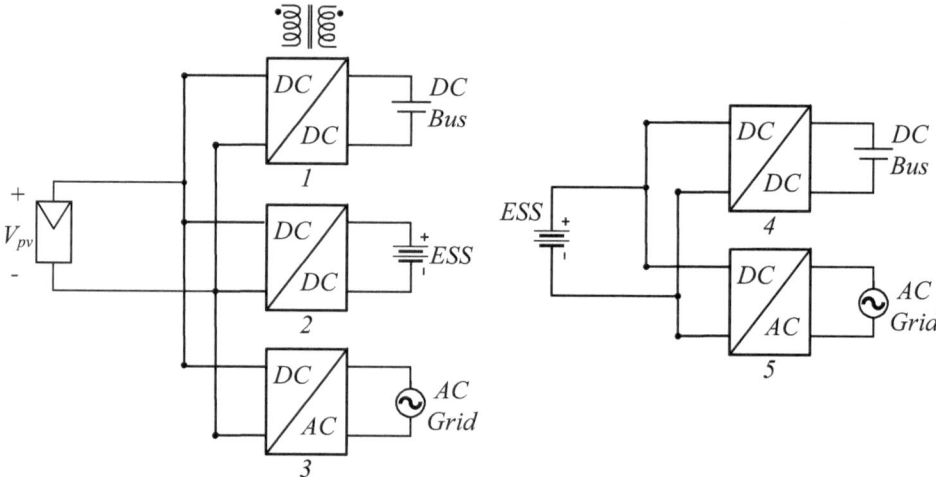

Fig. 1.7 Decentralized power conversion with multiple DAB converters interfacing PV, ESS, DC bus, and AC grid

References

1. W. Chen, F. C. Lee, M. M. Jovanovic, and J. A. Sabate, "A comparative study of a class of full bridge zero-voltage-power deviced PWM converters," in Proc. IEEE Appl. Power Electron. Conf. Expo., 1995, pp. 893–899.
2. G. Koo, G. Moon, and M. Youn, "New zero-voltage-switching phaseshift full-bridge converter with low conduction losses," IEEE Trans. Ind. Electron., vol. 52, no. 1, pp. 228–235, Feb. 2005.
3. Y. Jang and M. M. Jovanovic, "A new PWM ZVS full-bridge converter," IEEE Trans. Power Electron., vol. 22, no. 3, pp. 987–994, May 2007.
4. B. Zhang and S. Wang, "A Survey of EMI Research in Power Electronics Systems With Wide-Bandgap Semiconductor Devices," in IEEE Journal of Emerging and Selected Topics in Power Electronics, vol. 8, no. 1, pp. 626–643, March 2020, https://doi.org/10.1109/JESTPE.2019.2953730.
5. N. Jolly, A. Chandwani and A. Mallik, "Sliding Mode Control of a 2-MHz All-GaN-Based 700-W 95.6% Efficient LLC Converter," in IEEE Transactions on Transportation Electrification, vol. 9, no. 3, pp. 3677–3696, Sept. 2023, https://doi.org/10.1109/TTE.2022.3230929.
6. S. Mungekar and A. Mallik, "An Improved GHA-Enabled Steady State Model-Derived Semi-conductor Loss Optimization for a Three-Port C3L3 Resonant Converter," in IEEE Transactions on Power Electronics, vol. 39, no. 6, pp. 7654–7674, June 2024, https://doi.org/10.1109/TPEL.2024.3373514.
7. P. Purgat, S. Bandyopadhyay, Z. Qin and P. Bauer, "Zero Voltage Switching Criteria of Triple Active Bridge Converter," in *IEEE Transactions on Power Electronics*, vol. 36, no. 5, pp. 5425–5439, May 2021, https://doi.org/10.1109/TPEL.2020.3027785.
8. M. Rashidi, N. N. Altin, S. S. Ozdemir, A. Bani-Ahmed and A. Nasiri, "Design and Development of a High-Frequency Multiport Solid-State Transformer With Decoupled Control

Scheme," in *IEEE Transactions on Industry Applications*, vol. 55, no. 6, pp. 7515–7526, Nov.-Dec. 2019, https://doi.org/10.1109/TIA.2019.2939741.

9. J. Schäfer, D. Bortis and J. W. Kolar, "Multi-port multi-cell DC/DC converter topology for electric vehicle's power distribution networks," *2017 IEEE 18th Workshop on Control and Modeling for Power Electronics (COMPEL)*, 2017, pp. 1–9, https://doi.org/10.1109/COMPEL.2017.8013326.

10. G. Buticchi, L. F. Costa, D. Barater, M. Liserre and E. D. Amarillo, "A Quadruple Active Bridge Converter for the Storage Integration on the More Electric Aircraft," in *IEEE Transactions on Power Electronics*, vol. 33, no. 9, pp. 8174–8186, Sept. 2018, https://doi.org/10.1109/TPEL.2017.2781258.

Generalized Modeling Methodologies for Modular Multiport Power Converters

2.1 Introduction

The fundamental operating principle of all L-based non-resonant M2PC converters, including DAB, TAB, and higher-order MABs, is similar. The power transfer between any two ports of an M2PC primarily depends on the relative phase shifts between the voltages appearing across the corresponding switching cell outputs. By adjusting these relative phase shifts, the power flow can be regulated to the desired value [1–4].

Accurate modeling of M2PC converters is essential for understanding steady-state circuit waveforms, power flow, and other critical steady-state operational parameters such as winding current RMS, switch current RMS, and switching current peaks. These parameters provide valuable insights into the losses incurred by individual circuit elements, including MOSFETs, DC link capacitors, DC blocking capacitors, and magnetic components.

There are two primary approaches to modeling these converters: time-domain-oriented modeling [3, 5] and frequency-domain-based modeling [4, 6]. In the time-domain-oriented modeling technique, converter operation is divided into different operating modes based on the relative values of the phase shift control variables. In each mode, the inductor voltage and current take different waveforms. The circuit equations for each mode are solved separately to derive expressions for the inductor current, power transfer, and other relevant parameters. On the other hand, frequency-domain-oriented modeling offers a unified approach by decomposing phase-shift variable controlled converter's bridge voltages into a series of AC voltage harmonics with different amplitudes and frequencies using Fourier series expansion. This method bypasses the need for mode-dependent time-domain circuit analysis, which can become particularly complex in higher-order MABs. This approach simplifies the analysis and provides a comprehensive understanding of the converter's

© The Author(s), under exclusive license to Springer Nature Switzerland AG 2025   11
A. Mallik and S. Dey, *Switching Modulator Optimization in Isolated Power Converters*,
Synthesis Lectures on Power Electronics, https://doi.org/10.1007/978-3-031-81576-8_2

behavior across different operating conditions. Both methodologies are discussed in the following sections.

2.2 Segmentalized Time Domain Modeling of L-Based M2PC

The time-domain oriented modeling of the simplest L-based M2PC or the dual-active-bridge (DAB) converter and the triple-active-bridge (TAB) converter is presented in this section.

2.2.1 Time Domain Modeling of DAB

Figure 2.1 depicts the circuit topology of the dc-dc DAB converter. It comprises of two active H-bridges coupled via a high-frequency two-winding transformer with a turns ratio of $n_1 : n_2$. The series-connected inductors L_1 and L_2', located on the primary and secondary sides of the transformer windings, can be constituted by the leakage inductances of the transformer or by integrating external inductors in series with the transformer windings. The primary and secondary side active bridges are connected to dc link capacitors C_1 and C_2' with dc voltages V_1 and V_2' across them, respectively. Due to the controlled switching action of the MOSFETs in both H-bridges, two high-frequency AC voltages v_1 and v_2' emerge at the bridge outputs. The shapes of v_1 and v_2' are modulated by adjusting the phase shifts between the gating signals of the MOSFETs, thereby actively controlling the voltages across the inductors and resulting currents through the inductors. This control facilitates a specific power flow P_{12} between the two dc voltage sources V_1 and V_2'.

During modulation of the bridge output voltage shapes, three variables can be actively controlled:

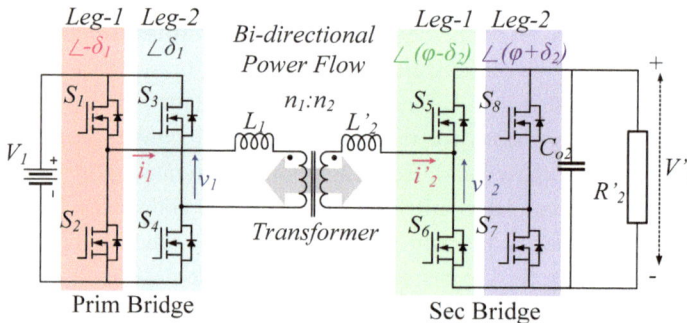

Fig. 2.1 DAB DC-DC Converter Topology and phase shifts of the individual half-bridge control signals

Fig. 2.2 Input and output side full-bridge voltages in relation with the gate signal phase displacements of the individual half-bridges. [T_s: switching period]

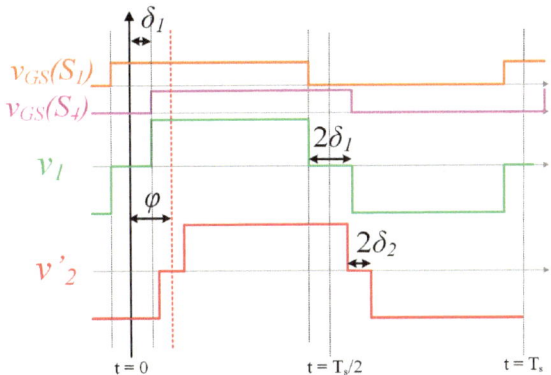

1. The inter-bridge phase shift (φ), which is the phase shift between fundamental components of v_1 and v_2'.
2. The intra-bridge phase shift of the primary H-bridge ($2\delta_1$), generated from the phase shift between the gating signals of S_1 and S_4, controlling the duty cycle of v_1.
3. The intra-bridge phase shift of the secondary H-bridge ($2\delta_2$), generated from the phase shift between the gating signals of S_5 and S_7, controlling the duty cycle of v_2'.

The relationship between these control variables and the gating signals of the primary DAB bridge, as well as the high-frequency ac voltage v_1, is depicted in Fig. 2.2. Utilizing these three phase-duty control variables, along with all gating signals having a 50% duty cycle and a switching frequency f_{sw}, a specific power flow can be achieved between the two dc ports of the DAB. The ranges of these DAB control variables are: $\varphi \in [-\pi/2, \pi/2]$ and $\delta_1, \delta_2 \in [0, \pi/2]$.

For ease of modeling, the DAB converter is represented as an equivalent circuit referred to the primary side of the transformer, as shown in Fig. 2.3. The electrical quantities in Fig. 2.3 can be expressed as follows:

$$L = L_1 + L_2; \quad L_2 = L_2'\left(\frac{n_1}{n_2}\right)^2; \quad v_2 = v_2'\left(\frac{n_1}{n_2}\right); \quad V_2 = V_2'\left(\frac{n_1}{n_2}\right) \tag{2.1}$$

In the DAB converter, the magnetizing inductance L_m is traditionally made significantly larger compared to L and thus, it carries very little current, not contributing to the power flow. Consequently, the magnetizing inductance L_m is not considered in the equivalent circuit.

Depending on the relative values of the three control variables that generate non-negative power transfer from port-1 to port-2 of the DAB (i.e., $P_{L12} > 0$, or, $0 \leq \delta_1, \delta_2, \varphi \leq \frac{\pi}{2}$), the operation of the converter can be categorized into five operating zones. As illustrated in Fig. 2.4, under each operating zone, the voltage across the equivalent circuit line inductor $v_L (= v_1 - v_2 = v_p - v_s)$ and the current through it $i_L (= i_1)$

Fig. 2.3 Equivalent circuit of the DAB converter

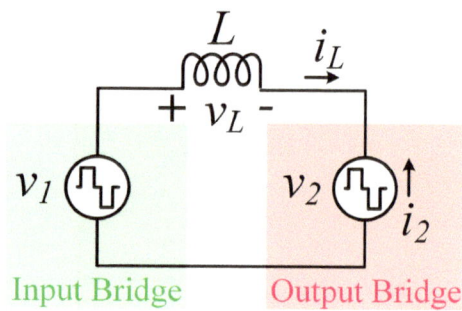

exhibit different profiles. The peaks of the inductor current i_L, denoted by $i_{p,1}$, $i_{p,2}$, $i_{s,1}$, and $i_{s,2}$ signifies the inductor current at the switching turn-on instant of the leading (leg-1) and lagging leg (leg-2) of the primary and secondary H-bridge, respectively. Moreover, by integrating the $i_L v_L$ product over a half switching cycle the average transferred power P_{12} in a DAB is determined:

$$P_{12} = \frac{2}{T_s} \int_0^{T_s/2} i_L v_L dt. \tag{2.2}$$

Each of the identified operating modes showcases different expressions for P_{12} as well as for instantaneous values of $i_L(t)$ during the switching instants for the leading and lagging switching legs of both the DAB H-bridges. The DAB circuit analysis under mode-1 is discussed below.

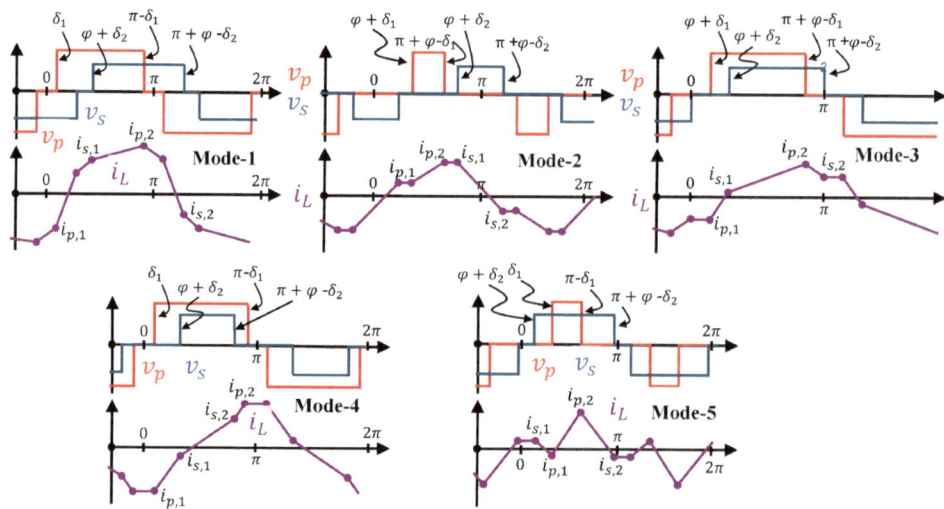

Fig. 2.4 Circuit waveforms of a DAB under different operating Modes for $\varphi > 0$

The condition for appearance of Mode-1 is given in (2.3).

$$\varphi \geq (\delta_1 + \delta_2); \ (\delta_1 + \delta_2) \leq \frac{\pi}{2} \tag{2.3}$$

Under this mode, the expressions of the normalized inductor current peaks (base current, $\frac{V_1}{X_L} = \frac{V_1}{2\pi f_{sw}L}$) can be derived applying volt-second balance across the inductor voltage v_L, and are given in (2.4).

$$\begin{cases} i_{p,1} = -[m\varphi - (1+m)\delta_1 + (1-m)\pi/2] \\ i_{p,2} = [m\varphi - (1-m)\delta_1 + (1-m)\pi/2] \\ i_{s,1} = [\varphi + (1-m)\delta_2 - (1-m)\pi/2] \\ i_{s,2} = -[\varphi - (1+m)\delta_2 - (1-m)\pi/2] \end{cases} \tag{2.4}$$

Here the output dc voltage gain of the converter is denoted as m ($= V_2/V_1$). It can be noticed from (2.4) that the direction and amplitude of the current at switching instants depend on the phase-duty control variables if other circuit conditions are kept unchanged. Moreover, these currents are vital to observe the soft-switching possibility of the DAB switching legs and to evaluate the switching losses in a DAB H-bridges.

Further, the expression for the per-unit power transfer ($P_{12(p.u.)}$) normalized with respect to base power (P_{base}) from primary to secondary side can be attained as,

$$P_{12(p.u.)} = \frac{1}{P_{base}} \frac{2}{T_s} \int_0^{T_s/2} i_L v_L dt = m \left[\varphi \left(1 - \frac{\varphi}{\pi} \right) - \frac{\delta_1^2 + \delta_2^2}{\pi} \right] \tag{2.5}$$

where the base power is, $P_{base} = \frac{V_1^2}{X_L} = \frac{V_1^2}{2\pi f_{sw}L}$. It is understandable from (2.5) that the amplitude of the output power in a DAB converter operating in Mode-1 is primarily dependent on the phase shift φ and the duty parameters δ_1 and δ_2, while the direction of the power flow entirely depends on φ. It is also noteworthy that the normalized maximum transfeable per unit power in a DAB converter is attained in this mode when $\delta_1 = \delta_2 = 0$ and $\varphi = \frac{\pi}{2}$; and can be written as (2.6).

$$P_{12, \max(p.u.)} = \frac{m\pi}{4} \tag{2.6}$$

The expressions of the DAB inductor current peaks and average power transfer under rest of the operating modes can be identified in a similar manner and is highlighted in Table 2.1.

The fundamental ZVS conditions for each HF switching leg in a DAB are primarily driven by the direction of the inductor current at the switching instants. According to the circuit waveforms presented in Fig. 2.4, the rising edge of v_p indicates the turn-on instant ($\theta = \delta_1$) for the high-side switch S_1 in leg-1 of primary bridge. Conversely, the falling edge of v_p signifies the turn-on instant ($\theta = \pi - \delta_1$) for the high-side switch S_3 in leg-2 of the same H-bridge. For the switch S_1 to achieve soft turn-on, the inductor current i_L

Table 2.1 Expressions of the DAB output power and current peaks under different operating modes

Operating modes	Condition	Turn-on switching currents of leading (1) and lagging leg (2) of the DAB ports (power flowing from port x to y) $[I_{base} = V_p^2/2\pi f_{sw}L]$		Transferred average Power P_{12} (p.u.); $P_{base} = V_p^2/2\pi f_{sw}L$
Mode-1	$\varphi \geq$ $\delta_1 + \delta_2$ $\delta_1+\delta_2 \leq \frac{\pi}{2}$	$i_{p,1}$	$-[m\varphi - (1+m)\delta_1 + (1-m)\pi/2]$	$m\left[\varphi\left(1 - \frac{\varphi}{\pi}\right) - \frac{\delta_1^2+\delta_2^2}{\pi}\right]$
		$i_{p,2}$	$[m\varphi - (1-m)\delta_1 + (1-m)\pi/2]$	
		$i_{s,1}$	$[\varphi + (1-m)\delta_2 - (1-m)\pi/2]$	
		$i_{s,2}$	$-[\varphi - (1+m)\delta_2 - (1-m)\pi/2]$	
Mode-2	$\frac{\pi}{2} \geq \varphi$ $\varphi \geq \pi - (\delta_1 + \delta_2)$	$i_{p,1}$	$-[m\delta_2 - \delta_1 + (1-m)\pi/2]$	$\frac{2m}{\pi}\left(\frac{\pi}{2} - \delta_1\right)\left(\frac{\pi}{2} - \delta_2\right)$
		$i_{p,2}$	$[-m\delta_2 - \delta_1 + (1+m)\pi/2]$	
		$i_{s,1}$	$[-m\delta_2 - \delta_1 + (1+m)\pi/2]$	
		$i_{s,2}$	$[m\delta_2 - \delta_1 + (1-m)\pi/2]$	
Mode-3	$|\delta_1 - \delta_2| \leq \varphi$ $\varphi < \min\begin{bmatrix}(\delta_1 + \delta_2), \\ (\pi - \delta_1 - \delta_2)\end{bmatrix}$	$i_{p,1}$	$-[m\delta_2 - \delta_1 + (1-m)\pi/2]$	$m\left[\varphi\left(1 - \frac{\varphi}{2\pi}\right) - \frac{\varphi(\delta_1+\delta_2)}{\pi} - \frac{(\delta_1-\delta_2)^2}{2\pi}\right]$
		$i_{p,2}$	$[m\varphi - (1-m)\delta_1 + (1-m)\pi/2]$	
		$i_{s,1}$	$[\varphi + (1-m)\delta_2 - (1-m)\pi/2]$	
		$i_{s,2}$	$[m\delta_2 - \delta_1 + (1-m)\pi/2]$	
Mode-4	$\varphi < |\delta_1-\delta_2|$ $\delta_2 > \delta_1$	$i_{p,1}$	$-[m\delta_2 - \delta_1 - (m-1)\pi/2]$	$m\varphi\left(1 - \frac{2\delta_2}{\pi}\right)$
		$i_{p,2}$	$[m\delta_2 - \delta_1 - (m-1)\pi/2]$	
		$i_{s,1}$	$[\varphi - (m-1)\delta_2 + (m-1)\pi/2]$	
		$i_{s,2}$	$[\varphi + (m-1)\delta_2 - (m-1)\pi/2]$	

(continued)

Table 2.1 (continued)

Operating modes	Condition	Turn-on switching currents of leading (1) and lagging leg (2) of the DAB ports (power flowing from port x to y) $[I_{base} = V_p^2/2\pi f_{sw}L]$		Transferred average Power P_{12} (p.u.); $P_{base} = V_p^2/2\pi f_{sw}L$
Mode-5	$\varphi < \|\delta_1 - \delta_2\|$	$i_{p,1}$	$-[-m\varphi - (1-m)\delta_1 + (1-m)\pi/2]$	$m\varphi\left(1 - \frac{2\delta_1}{\pi}\right)$
	$\delta_1 > \delta_2$	$i_{p,2}$	$[m\varphi - (1-m)\delta_1 + (1-m)\pi/2]$	
		$i_{s,1}$	$[\delta_1 - m\delta_2 - (1-m)\pi/2]$	
		$i_{s,2}$	$-[\delta_1 - m\delta_2 - (1-m)\pi/2]$	

at $\theta = \delta_1$ ($i_L|_{\theta=\delta_1} = i_{p,1}$) must be directed towards primary bridge (i.e., $i_{p,1} < 0$). This condition ensures that the body capacitor of S_1 (denoted as C_{oss,S_1}) will discharge while the body capacitor of S_2 charges during the deadtime before S_1 turns on. As a result, S_1 can turn on with zero voltage across the device ($v_{DS,S_1} = 0$). Thus, if judged solely based on the inductor current direction, the ZVS conditions for the four HF switching legs in the DAB can be summarized as follows:

$$i_{p,1}\langle 0; \ i_{p,2}\rangle 0; \ i_{s,1}\rangle 0; \ i_{s,2}\langle 0. \tag{2.7}$$

By examining the expressions for $i_{p,1}, i_{p,2}, i_{s,1}$, and $i_{s,2}$ under the five modes of DAB operation highlighted in Table 2.1, it can be concluded that ZVS conditions are simultaneously satisfied only in mode-1, mode-4, and mode-5. In contrast, for modes 2 and 3, the leading leg of the primary H-bridge and the lagging leg of the secondary H-bridge cannot achieve soft-switching simultaneously because, in both these modes, the relation $i_{p,1} = -i_{s,2}$ violates the conditions outlined in (2.7).

2.2.2 Time Domain Modeling of a TAB

The TAB converter is a DAB-derived topology where an additional H-bridge is coupled with two active bridges present in a DAB, utilizing a three-winding HF transformer. Figure 2.5 illustrates the schematic of the Triple Active Bridge (TAB) converter, which features three DC voltage ports, i.e., sources or loads denoted as V_1, V_2', and V_3'. These sources are connected to a three-terminal transformer via three full-bridge cells. The transformer functions as an AC-link, coupling these ports at distinct voltage levels determined by the corresponding turn ratios $n_1 : n_2 : n_3$. The inductors L_1, L_2' and L_3' can either be separate magnetic components or derived from the transformer's leakage inductances.

The phase shifts between the gate driving signals of all high-frequency switching legs in a TAB result in the H-bridge output voltages v_1, $v_{2'}$ and $v_{3'}$, as illustrated in Fig. 2.6. The fundamental components of $v_{2'}$ and $v_{3'}$ are phase-shifted by φ_2 and φ_3 relative to v_1. Traditionally, inter-bridge phase shifts (φ_2 and φ_3) have been the primary means of controlling power flow in a TAB converter. However, introducing intra-bridge phase shifts ($2\delta_1, 2\delta_2$ and $2\delta_3$) provides enhanced control over power flow with greater degrees of freedom involved. The control variable ranges for the TAB are $\varphi_k \in [-\pi/2, \pi/2]$ and $\delta_k \in [0, \pi/2]$, where $k = 1, 2$ or 3 and $\varphi_1 = 0$. The average power sourced by each of the dc TAB ports are denoted as P_1, P_2, and P_3.

Figure 2.7 presents a simplified schematic of the TAB network, transformed into primary-referred Y and followed by a Δ-equivalent circuit model. The electrical quantities of the Y-equivalent can be expressed as follows:

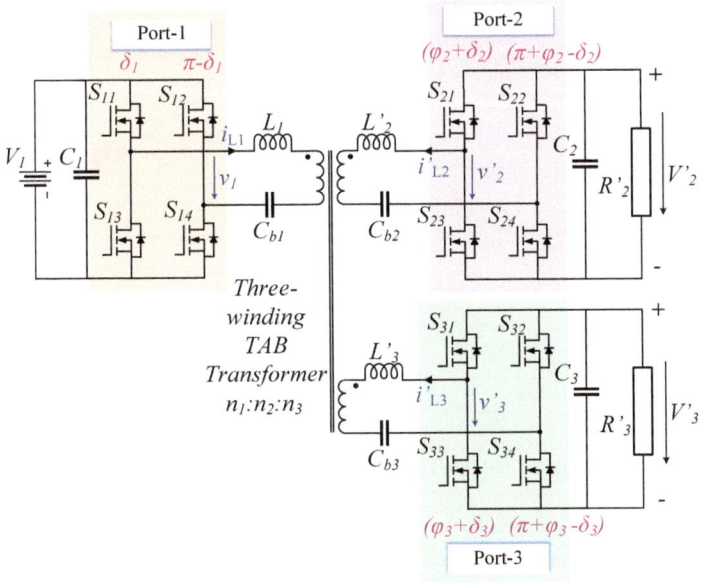

Fig. 2.5 TAB converter topology and phase shifts of the individual half-bridge gating signals

Fig. 2.6 Full-bridge output voltages in relation with the gate signal phase displacements of the individual half-bridges

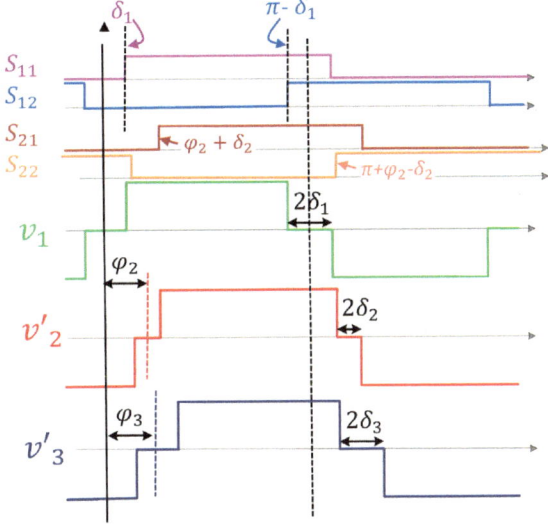

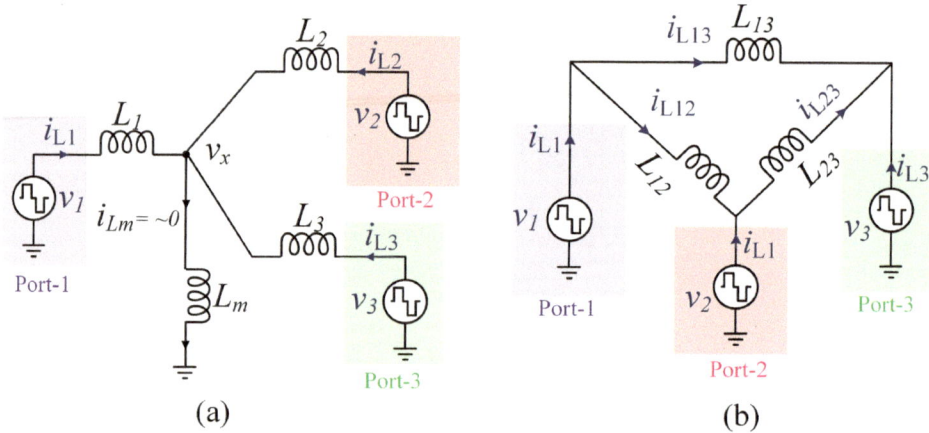

Fig. 2.7 Equivalent circuits of the TAB converter: **a** Y-equivalent and **b** Δ-equivalent circuit

$$\begin{cases} L_2 = L_2'\left(\frac{n_1}{n_2}\right)^2; \ v_2 = v_2'\left(\frac{n_1}{n_2}\right); \ i_2 = i_2'\left(\frac{n_2}{n_1}\right) \\ L_3 = L_3'\left(\frac{n_1}{n_3}\right)^2; \ v_3 = v_3'\left(\frac{n_1}{n_3}\right); \ i_3 = i_3'\left(\frac{n_3}{n_1}\right). \end{cases} \tag{2.8}$$

The Δ-equivalent circuit inductances are formed using Y-Δ transformation laws and can be represented as (2.9).

$$\begin{aligned} L_{12} &= L_1 + L_2 + L_1 L_2 / L_3 \\ L_{13} &= L_1 + L_3 + L_1 L_3 / L_2 \\ L_{23} &= L_2 + L_3 + L_2 L_3 / L_1. \end{aligned} \tag{2.9}$$

In accordance with the Δ circuit model, the total instantaneous current in each TAB port can be divided in two parts using Kirchhoff's current law (KCL) and can form the following relationships:

$$\begin{cases} i_{L1}(t) = i_{L12}(t) + i_{L13}(t) \\ i_{L2}(t) = i_{L23}(t) - i_{L12}(t) \ . \\ i_{L3}(t) = -i_{L23}(t) - i_{L13}(t) \end{cases} \tag{2.10}$$

Further, the average power sourced by each TAB dc port can be expressed as follows.

$$\begin{cases} P_1 = P_{L12} + P_{L13} \\ P_2 = P_{L23} - P_{L12} \ . \\ P_3 = -P_{L23} - P_{L13} \end{cases} \tag{2.11}$$

Here, i_{Lxy} and P_{Lxy} denote the current and average power flowing through the Δ-model HF inductor (L_{xy}), from port- x to y. In the context of Δ-model inductors connected to

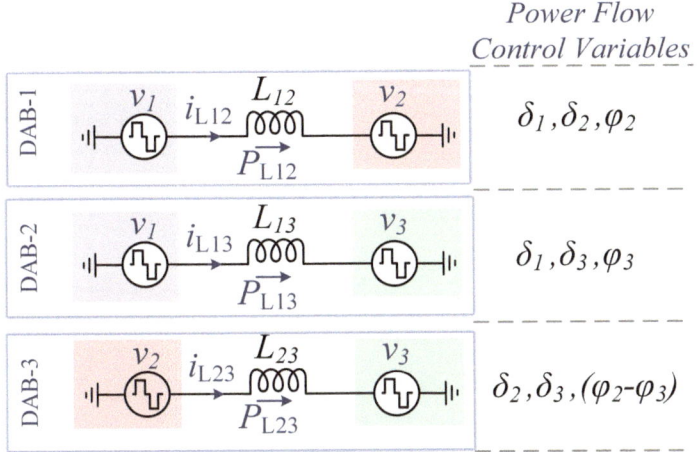

Fig. 2.8 DAB-based equivalent circuit representation of the Δ-TAB circuit

each TAB port, the current flowing through each port's inductor can be viewed as the sum of two branch currents, each influenced solely by the voltage sources connected to the respective ends of the inductor. The current i_{Lxy} and power P_{Lxy} flowing from port x to port y through L_{xy} depend exclusively on the intra-bridge phase shifts of ports x and y (δ_x and δ_y) and the relative phase difference between the two corresponding ports ($\varphi_y - \varphi_x$). Consequently, the TAB converter can be decomposed into three separate DAB converters, as illustrated in Fig. 2.8.

To synthesize the TAB bridge currents, voltages, and power flow among the ports, the mode-dependent time-domain circuit analysis of a standalone DAB cell, presented in Sect. 2.2.1, needs to be utilized. This analysis provides the foundational principles required to understand and model the behavior of the TAB converter using the DAB knowledgebase. For a standalone xy-DAB cell with port-x and port-y with bridge output voltages v_x and v_y and a line inductor L_{xy} connected in between, the derived expressions of the power transfer P_{Lxy} and instantaneous inductor current at the switching instants, i.e., $i_{x,1}$, $i_{x,2}$, $i_{y,1}$, and $i_{y,2}$ are highlighted in Table 2.2 for five different possible DAB operating modes.

Once the standalone DAB converter's analytical modeling is complete, the winding currents and power flow of the TAB converter can be accurately quantified by employing the superposition theorem, stated in (2.10) and (2.11). In order to use the DAB peak current and power flow expressions to compute the TAB currents, the direction of the power flow within the Δ-circuit DAB cells needs to be identified first. Any TAB port is common to two separate DAB cells and therefore, whether that port is sourcing or sinking power, four different combinations of DAB power flow can appear that leads

Table 2.2 Switching currents and power flow analysis in a xy-DAB cell under different operating modes

Operating modes	Condition	Turn-on switching currents of leading (1) and lagging leg (2) of the DAB ports (power flowing from port x to y) $[pu_{xy} = V_x/2\pi f_{sw} L_{xy}]$		Transferred average Power P_{Lxy} (p.u.); $P_{base} = V_x^2/2\pi f_{sw} L_{xy}$
Mode-1	$(\varphi_y - \varphi_x) \geq (\delta_x + \delta_y)$; $(\delta_x + \delta_y) \leq \frac{\pi}{2}$	$i_{x,1}$	$-pu_{xy}\left[m_{xy}(\varphi_y - \varphi_x) - (1 + m_{xy})\delta_x + (1 - m_{xy})\pi/2\right]$	$m_{xy}\left[(\varphi_y - \varphi_x)\left\{1 - \frac{(\varphi_y - \varphi_x)}{\pi}\right\} - \frac{\delta_x^2 + \delta_y^2}{\pi}\right]$
		$i_{x,2}$	$pu_{xy}\left[m_{xy}(\varphi_y - \varphi_x) - (1 - m_{xy})\delta_x + (1 - m_{xy})\pi/2\right]$	
		$i_{y,1}$	$pu_{xy}\left[(\varphi_y - \varphi_x) + (1 - m_{xy})\delta_y - (1 - m_{xy})\pi/2\right]$	
		$i_{y,2}$	$-pu_{xy}\left[(\varphi_y - \varphi_x) - (1 + m_{xy})\delta_y - (1 - m_{xy})\pi/2\right]$	
Mode-2	$\frac{\pi}{2} \geq (\varphi_y - \varphi_x)$; $(\varphi_y - \varphi_x) \geq \pi - (\delta_x + \delta_y)$	$i_{x,1}$	$-pu_{xy}\left[m_{xy}\delta_y - \delta_x + (1 - m_{xy})\pi/2\right]$	$\frac{2m_{xy}}{\pi}\left(\frac{\pi}{2} - \delta_x\right)\left(\frac{\pi}{2} - \delta_y\right)$
		$i_{x,2}$	$pu_{xy}\left[-m_{xy}\delta_y - \delta_x + (1 + m_{xy})\pi/2\right]$	
		$i_{y,1}$	$pu_{xy}\left[-m_{xy}\delta_y - \delta_x + (1 + m_{xy})\pi/2\right]$	
		$i_{y,2}$	$pu_{xy}\left[m_{xy}\delta_y - \delta_x + (1 - m_{xy})\pi/2\right]$	
Mode-3	$|\delta_x - \delta_y| \leq (\varphi_y - \varphi_x)$; $(\varphi_y - \varphi_x) < \min\left[(\delta_x + \delta_y), (\pi - \delta_x - \delta_y)\right]$	$i_{x,1}$	$-pu_{xy}\left[m_{xy}(\varphi_y - \varphi_x) - (1 - m_{xy})\delta_x + (1 - m_{xy})\pi/2\right]$	$m_{xy}\left[(\varphi_y - \varphi_x)\left(1 - \frac{\varphi_y - \varphi_x}{2\pi}\right) - \frac{(\varphi_y - \varphi_x)(\delta_x + \delta_y)}{\pi} - \frac{(\delta_x - \delta_y)^2}{2\pi}\right]$
		$i_{x,2}$	$pu_{xy}\left[m_{xy}(\varphi_y - \varphi_x) - (1 - m_{xy})\delta_x + (1 - m_{xy})\pi/2\right]$	
		$i_{y,1}$	$pu_{xy}\left[(\varphi_y - \varphi_x) + (1 - m_{xy})\delta_y - (1 - m_{xy})\pi/2\right]$	
		$i_{y,2}$	$pu_{xy}\left[m_{xy}\delta_y - \delta_x + (1 - m_{xy})\pi/2\right]$	
Mode-4	$(\varphi_y - \varphi_x) < |\delta_x - \delta_y|$; $\delta_y > \delta_x$	$i_{x,1}$	$-pu_{xy}\left[m_{xy}(\varphi_y - \varphi_x) - \delta_x + (1 - m_{xy})\pi/2\right]$	$m_{xy}(\varphi_y - \varphi_x)\left(1 - \frac{2\delta_y}{\pi}\right)$
		$i_{x,2}$	$pu_{xy}\left[m_{xy}\delta_x - \delta_x + (1 - m_{xy})\pi/2\right]$	
		$i_{y,1}$	$pu_{xy}\left[(\varphi_y - \varphi_x) + (1 - m_{xy})\delta_y - (1 - m_{xy})\pi/2\right]$	
		$i_{y,2}$	$pu_{xy}\left[(\varphi_y - \varphi_x) - (1 - m_{xy})\delta_y + (1 - m_{xy})\pi/2\right]$	
Mode-5	$(\varphi_y - \varphi_x) < |\delta_x - \delta_y|$; $\delta_x > \delta_y$	$i_{x,1}$	$-pu_{xy}\left[-m_{xy}(\varphi_y - \varphi_x) - (1 - m_{xy})\delta_x + (1 - m_{xy})\pi/2\right]$	$m_{xy}(\varphi_y - \varphi_x)\left(1 - \frac{2\delta_x}{\pi}\right)$
		$i_{x,2}$	$pu_{xy}\left[m_{xy}(\varphi_y - \varphi_x) - (1 - m_{xy})\delta_x + (1 - m_{xy})\pi/2\right]$	
		$i_{y,1}$	$pu_{xy}\left[\delta_x - m_{xy}\delta_y - (1 - m_{xy})\pi/2\right]$	
		$i_{y,2}$	$-pu_{xy}\left[\delta_x - m_{xy}\delta_y - (1 - m_{xy})\pi/2\right]$	

to four different current and power flow expressions for that TAB port. The TAB port-1 bridge current $i_{L1}(t)$ at the turn-on transient of leading (leg 1) and lagging (leg 2) switching legs are deduced as $i_{L1,1}$ and $i_{L1,2}$, with their respective expressions outlined in (2.12) and (2.13).

$$
i_{L1,1} = \begin{cases} i_{1,1}\big|_{P_{L12}} + i_{1,1}\big|_{P_{L13}}, & if\ \varphi_2 > \varphi_1\ or\ P_{12} > 0\ and\ \varphi_3 > \varphi_1\ or\ P_{13} > 0 \\ i_{1,1}\big|_{P_{L12}} - i_{1,1}\big|_{P_{L31}}, & if\ \varphi_2 > \varphi_1\ or\ P_{12} > 0\ and\ \varphi_3 < \varphi_1\ or\ P_{13} < 0 \\ -i_{1,1}\big|_{P_{L21}} - i_{1,1}\big|_{P_{L31}}, & if\ \varphi_2 < \varphi_1\ or\ P_{12} < 0\ and\ \varphi_3 < \varphi_1\ or\ P_{13} < 0 \\ -i_{1,1}\big|_{P_{L21}} + i_{1,1}\big|_{P_{L13}}, & if\ \varphi_2 < \varphi_1\ or\ P_{12} < 0\ and\ \varphi_3 > \varphi_1\ or\ P_{13} > 0 \end{cases} \quad (2.12)
$$

$$
i_{L1,2} = \begin{cases} i_{1,2}\big|_{P_{L12}} + i_{1,2}\big|_{P_{L13}}, & if\ \varphi_2 > \varphi_1\ or\ P_{12} > 0\ and\ \varphi_3 > \varphi_1\ or\ P_{13} > 0 \\ i_{1,2}\big|_{P_{L12}} - i_{1,2}\big|_{P_{L31}}, & if\ \varphi_2 > \varphi_1\ or\ P_{12} > 0\ and\ \varphi_3 < \varphi_1\ or\ P_{13} < 0 \\ -i_{1,2}\big|_{P_{L21}} - i_{1,2}\big|_{P_{L31}}, & if\ \varphi_2 < \varphi_1\ or\ P_{12} < 0\ and\ \varphi_3 < \varphi_1\ or\ P_{13} < 0 \\ -i_{1,2}\big|_{P_{L21}} + i_{1,2}\big|_{P_{L13}}, & if\ \varphi_2 < \varphi_1\ or\ P_{12} < 0\ and\ \varphi_3 > \varphi_1\ or\ P_{13} > 0 \end{cases} \quad (2.13)
$$

where $i_{1,1}\big|_{P_{L12},x=1,y=2}$ and $i_{1,2}\big|_{P_{L12},x=1,y=2}$ are the L_{12} inductor current (i_{L12}) during turn-on instants of the leading and lagging legs of port-1 H-bridge inside DAB-12 module (as given in Fig. 2.8), respectively, and the active power is being transferred from port-1 to port-2 or $P_{L12} > 0$. If power is transferred in opposite direction, i.e., $P_{L12} < 0$, the turn-on switching current of the leading and lagging legs of port-1 are given as $i_{1,1}\big|_{P_{L21},x=2,y=1}$ and $i_{1,2}\big|_{P_{L21},x=2,y=1}$, correspondingly. Similarly, $i_{1,1}\big|_{P_{L13},x=1,y=3}$ and $i_{1,2}\big|_{P_{L13},x=1,y=3}$ represent the instantaneous values of i_{L13} during turn-on instants of port-1 H-bridge leading and lagging switching legs (DAB-13), respectively, while $P_{L13} > 0$. If $P_{L13} < 0$, these two currents are depicted by $i_{1,1}\big|_{P_{L31},x=3,y=1}$ and $i_{1,2}\big|_{P_{L31},x=3,y=1}$, correspondingly. The values of $i_{1,1}\big|_{P_{L12}}$, $i_{1,1}\big|_{P_{L21}}$, $i_{1,1}\big|_{P_{L13}}$, $i_{1,1}\big|_{P_{L31}}$, $i_{1,2}\big|_{P_{L12}}$, $i_{1,2}\big|_{P_{L21}}$, $i_{1,2}\big|_{P_{L13}}$ and $i_{1,2}\big|_{P_{L31}}$ can be calculated based on the operating modes of the corresponding DAB cells from Table 2.2.

In a similar manner the rest two winding currents of the TAB ($i_{L2}(t)$ and $i_{L3}(t)$) during the turn-on instants of port-2 and port-3 switching legs can be synthesized using the superposition theorem of (2.10) and are highlighted as $i_{L2,1}$, $i_{L2,2}$ and $i_{L3,1}$, $i_{L3,2}$, respectively, in (2.14) and (2.15).

$$
\begin{cases} i_{L2,1} = \begin{cases} -i_{2,1}\big|_{P_{L12},x=1,y=2} + i_{2,1}\big|_{P_{L23},x=2,y=3}, & if\ \varphi_3 > \varphi_2\ or\ P_{23} > 0\ and\ \varphi_2 > \varphi_1\ or\ P_{12} > 0 \\ -i_{2,1}\big|_{P_{L12},x=1,y=2} - i_{2,1}\big|_{P_{L32},x=3,y=2}, & if\ \varphi_3 < \varphi_2\ or\ P_{23} < 0\ and\ \varphi_2 > \varphi_1\ or\ P_{12} > 0 \\ i_{2,1}\big|_{P_{L21},x=2,y=1} - i_{2,1}\big|_{P_{L32},x=3,y=2}, & if\ \varphi_3 < \varphi_2\ or\ P_{23} < 0\ and\ \varphi_2 < \varphi_1\ or\ P_{12} < 0 \\ i_{2,1}\big|_{P_{L21},x=2,y=1} + i_{2,1}\big|_{P_{L23},x=2,y=3}, & if\ \varphi_3 > \varphi_2\ or\ P_{23} > 0\ and\ \varphi_2 < \varphi_1\ or\ P_{12} < 0 \end{cases} \\[2em] i_{L2,2} = \begin{cases} -i_{2,2}\big|_{P_{L12},x=1,y=2} + i_{2,2}\big|_{P_{L23},x=2,y=3}, & if\ \varphi_3 > \varphi_2\ or\ P_{23} > 0\ and\ \varphi_2 > \varphi_1\ or\ P_{12} > 0 \\ -i_{2,2}\big|_{P_{L12},x=1,y=2} - i_{2,2}\big|_{P_{L32},x=3,y=2}, & if\ \varphi_3 < \varphi_2\ or\ P_{23} < 0\ and\ \varphi_2 > \varphi_1\ or\ P_{12} > 0 \\ i_{2,2}\big|_{P_{L21},x=2,y=1} - i_{2,2}\big|_{P_{L32},x=3,y=2}, & if\ \varphi_3 < \varphi_2\ or\ P_{23} < 0\ and\ \varphi_2 < \varphi_1\ or\ P_{12} < 0 \\ i_{2,2}\big|_{P_{L21},x=2,y=1} + i_{2,2}\big|_{P_{L23},x=2,y=3}, & if\ \varphi_3 > \varphi_2\ or\ P_{23} > 0\ and\ \varphi_2 < \varphi_1\ or\ P_{12} < 0 \end{cases} \end{cases}
$$

$$(2.14)$$

$$
\begin{cases}
i_{L3,1} = \begin{cases}
-i_{3,1}\big|_{P_{L13},x=1,y=3} - i_{3,1}\big|_{P_{L23},x=2,y=3}, & \text{if } \varphi_3 > \varphi_2 \text{ or } P_{23} > 0 \text{ and } \varphi_3 > \varphi_1 \text{ or } P_{13} > 0 \\
-i_{3,1}\big|_{P_{L13},x=1,y=3} + i_{3,1}\big|_{P_{L32},x=3,y=2}, & \text{if } \varphi_3 < \varphi_2 \text{ or } P_{23} < 0 \text{ and } \varphi_3 > \varphi_1 \text{ or } P_{13} > 0 \\
i_{3,1}\big|_{P_{L31},x=3,y=1} - i_{3,1}\big|_{P_{L23},x=2,y=3}, & \text{if } \varphi_3 > \varphi_2 \text{ or } P_{23} > 0 \text{ and } \varphi_3 < \varphi_1 \text{ or } P_{13} < 0 \\
i_{3,1}\big|_{P_{L31},x=3,y=1} + i_{3,1}\big|_{P_{L32},x=3,y=2}, & \text{if } \varphi_3 < \varphi_2 \text{ or } P_{23} < 0 \text{ and } \varphi_3 < \varphi_1 \text{ or } P_{13} < 0
\end{cases} \\
i_{L3,2} = \begin{cases}
-i_{3,2}\big|_{P_{L13},x=1,y=3} - i_{3,2}\big|_{P_{L23},x=2,y=3}, & \text{if } \varphi_3 > \varphi_2 \text{ or } P_{23} > 0 \text{ and } \varphi_3 > \varphi_1 \text{ or } P_{13} > 0 \\
-i_{3,2}\big|_{P_{L13},x=1,y=3} + i_{3,2}\big|_{P_{L32},x=3,y=2}, & \text{if } \varphi_3 < \varphi_2 \text{ or } P_{23} < 0 \text{ and } \varphi_3 > \varphi_1 \text{ or } P_{13} > 0 \\
i_{3,2}\big|_{P_{L31},x=3,y=1} - i_{3,2}\big|_{P_{L23},x=2,y=3}, & \text{if } \varphi_3 > \varphi_2 \text{ or } P_{23} > 0 \text{ and } \varphi_3 < \varphi_1 \text{ or } P_{13} < 0 \\
i_{3,2}\big|_{P_{L31},x=3,y=1} + i_{3,2}\big|_{P_{L32},x=3,y=2}, & \text{if } \varphi_3 < \varphi_2 \text{ or } P_{23} < 0 \text{ and } \varphi_3 < \varphi_1 \text{ or } P_{13} < 0
\end{cases}
\end{cases}
\tag{2.15}
$$

The average power sourced by each of the TAB ports can be computed using (2.11), with the given relative values of the phase-shift control variables and the power transfer expressions of the DAB cells under the corresponding modes of operations given in Table 2.2.

2.2.3 Time Domain Modeling of a n-Port MAB

The time domain modeling approach for a TAB converter using the superposition principle and fundamental DAB cell voltage and current expressions can be extrapolated to a generic n-port MAB converter. Figure 2.9 depicts the schematic of a generic n-port MAB converter, where multiple active full bridges (ABs) are magnetically coupled through an n-winding HF transformer with turns ratios $n_1, n_2, \ldots n_n$. Each port-x ($1 \leq x \leq n$) includes a dc blocking capacitor C_x in series with its line inductance L'_x to mitigate transformer bias issues during transient events such as load changes. These capacitors are sized large enough to provide significantly lower impedance compared to the inductive impedance, resulting in a resonant frequency with L'_x that is less than one-tenth of the switching frequency, thus ensuring they do not interfere with the power flow in the MAB converter. Similar to any DAB and TAB converter, the control of power flow within the MAB converter is governed by the bridge voltage duty ratios (δ_x) and the phase differences between the active bridges (φ_x, where $1 \leq x \leq n$ and $\varphi_1 = 0$). The phase displacement between the gating signals of the two half-bridge (HB) legs of any specific AB determines the duty cycle of the bridge's output voltage, as illustrated in Fig. 2.10.

A simplified star-equivalent or Y-model is used to analyze the operation of the MAB converter that is presented in Fig. 2.11a. For ease of analysis all the generated ac voltage sources ($v_{x'}$) and the transformer leakage inductances ($L_{x'}$) are referred to the port 1. Thus, the modified circuit parameters are represented as:

$$
L_x = L'_x \left(\frac{n_x}{n_1}\right)^2; \quad v_x = v'_x \left(\frac{n_1}{n_x}\right); \quad i_x = i'_x \left(\frac{n_x}{n_1}\right); \quad V_x = V'_x \left(\frac{n_1}{n_x}\right).
\tag{2.16}
$$

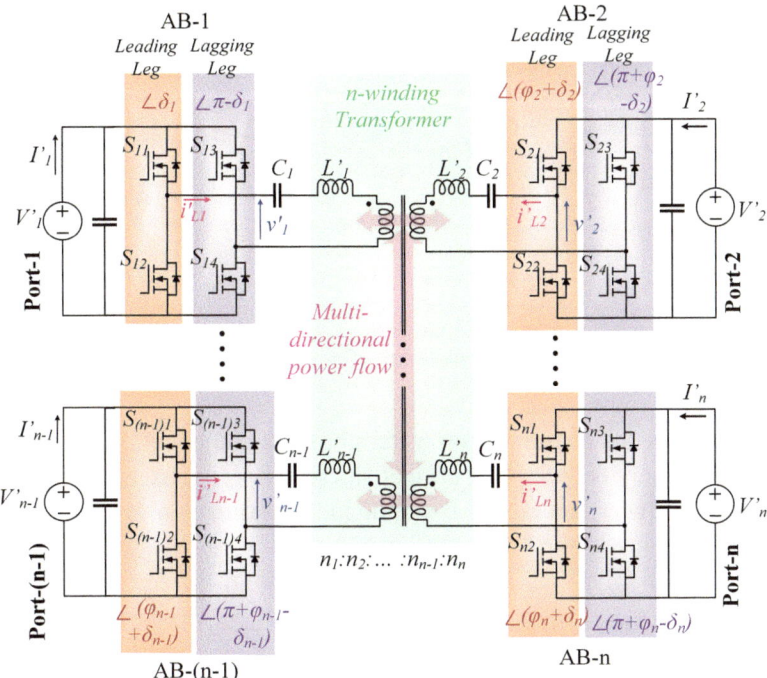

Fig. 2.9 A generic n-port MAB converter topology and phase shifts of the respective half-bridge gating signals

Fig. 2.10 MAB Full-bridge output voltages in relation with the gate signal phase displacements of the individual half-bridges

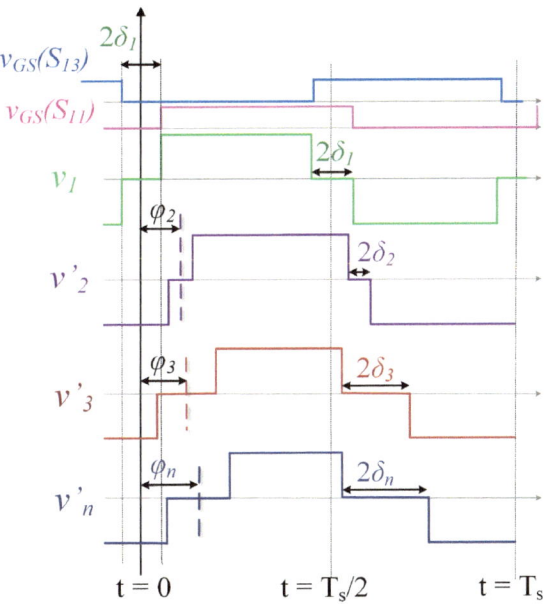

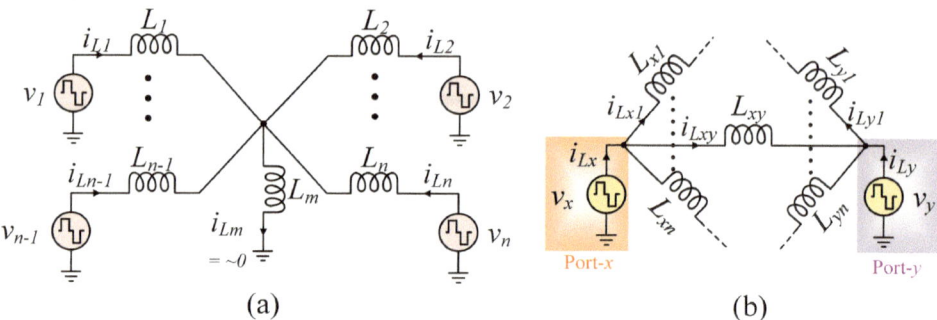

Fig. 2.11 **a** Y-equivalent; **b** Δ-equivalent circuit of n-port MAB

In any DAB-derived converter, power transfer through the line inductance between the ports is possible only between two distinct sinusoidal port voltage sources that have the same frequency. Under mode-1 DAB operation, the expression of the power flow from port-x to port-y is given by,

$$P_{xy} = \frac{4V_x V_y}{\pi^3 f_{sw} L_{xy}} \cdot \left[(\varphi_y - \varphi_x)\left(1 - \frac{\varphi_y - \varphi_x}{\pi}\right) - \frac{\delta_x^2 + \delta_y^2}{\pi} \right] \qquad (2.17)$$

where φ_x and φ_y are the corresponding phase shifts; V_x and V_y are the dc link voltages, respectively and L_{xy} is the equivalent line inductance between the ports- x and y. Therefore, to utilize the power flow Eq. (2.17) in formulating the power flow in a n-port MAB converter, the quantification of the inter port line inductance, L_{xy}, is essential. This leads to the formation of a Δ model of the n-port MAB structure (see Fig. 2.11b) where the newly formed interport-inductors connect the ac port voltage sources with a total number of links of $C_2^n = \frac{n!}{2 \bullet (n-2)!}$. Theory of superposition is applied on the Y-model (Fig. 2.12a) in order to obtain the Δ model of the converter. To calculate the power flow from port-x to port-y in the Y-model, all the other voltage sources are kept short circuited that results in the equivalent circuit, displayed in Fig. 2.12b. The Thevenin-equivalent inductance between the star point and ground and the Thevenin-equivalent ac voltage of the star point can be calculated as,

$$L_{TH,star} = \left(\frac{1}{L_m} + \sum_{k \neq x,y}^{n} \frac{1}{L_k} \right)^{-1} \cong \left(\sum_{k \neq x,y}^{n} \frac{1}{L_k} \right)^{-1} \qquad (2.18)$$

$$v_{TH,star} = \left(\frac{L_x \| L_{TH,star}}{L_y + L_x \| L_{TH,star}} \right) v_y + \left(\frac{L_y \| L_{TH,star}}{L_x + L_y \| L_{TH,star}} \right) v_x \qquad (2.19)$$

Now the power transfer from port-x to port-y can be seen as equivalent to the power transfer between port-x and the $v_{TH,star}$ through the line inductor L_x (Fig. 2.12). Also,

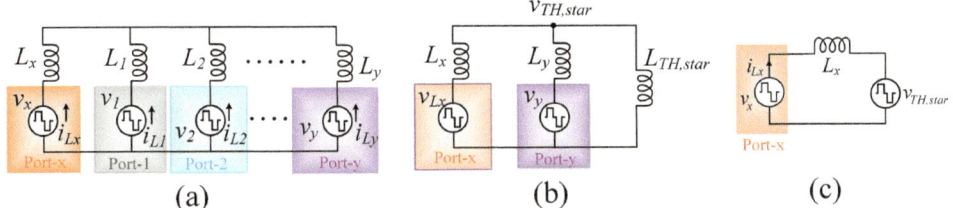

Fig. 2.12 Derivation of the inter-port Δ-model MAB Inductance between port-x and y: **a** MAB Y-model; **b** Thevenin-equivalent impedance and voltage at star-point with rest port voltages shorted; **c** Obtained Thevenin-equivalent circuit with respect to port-x

while computing the power flow between these two nodes, the contribution of the v_x in $v_{TH,star}$ does not generate any active power flow as it does not contain any phase shift with respect to the xth port voltage. Thus, substituting $v_{TH,star}$ with only the contribution of v_y in (2.17), the generalized power flow between two MAB ports is determined as,

$$P_{xy} = P_{xv_{th}} = \frac{4V_x V_y \left(\frac{L_x \| L_{TH,star}}{L_y + L_x \| L_{TH,star}} \right)}{\pi^3 f_{sw} L_x} \cdot \left[(\varphi_y - \varphi_x)\left(1 - \frac{\varphi_y - \varphi_x}{\pi}\right) - \frac{\delta_x^2 + \delta_y^2}{\pi} \right] \quad (2.20)$$

Therefore, comparing (2.20) with (2.17), the equivalent inter-port link inductance between the MAB ports x and y is formulated as,

$$L_{xy} = L_x + L_y + \frac{L_x L_y}{L_{TH,star}} = L_x + L_y + L_x L_y \left(\sum_{k \neq x,y}^{n} \frac{1}{L_k} \right); x \neq y \quad (2.21)$$

Now, applying KCL on the Δ-model of the MAB converter, the total instantaneous current in each MAB port (port-x) and the average power sourced by that port can be represented as summation of the currents and powers flowing from port-x to any other MAB port through the inter-port line inductances (L_{xy}) and are highlighted in (2.22) and (2.23), respectively.

$$i_{Lx}(t) = \sum_{\substack{y=1 \\ y \neq x}}^{n} i_{Lxy}(t). \quad (2.22)$$

$$P_x = \sum_{\substack{y=1 \\ y \neq x}}^{n} P_{Lxy}. \quad (2.23)$$

Thus, the MAB Δ-model can be seen as a superposition of C_2^n individual DAB cells, which we noticed in case of time domain TAB circuit modeling as well. Based on the power flow direction concerning a MAB port, i.e., if the port acts like a power load or source based each of (n-1) DAB cells it is connected to, the peak switching currents of that MAB port during turn-on instants can be obtained. Consolidating the expressions of turn-on switching currents for a TAB converter given in (2.12)–(2.15), the HF inductor current i_{Lx} corresponding to any port-j (of Y-model) of a n-port MAB converter during its leading and lagging switching leg turn-on instants can be identified using the unified expression shown in (2.24).

$$
i_{Lx,j} = \sum_{\substack{y=1 \\ y \neq x}}^{n} \begin{cases} i_{x,j}\big|_{P_{xy}}, & \text{if } \varphi_y > \varphi_x \\ -i_{x,j}\big|_{P_{yx}}, & \text{if } \varphi_y < \varphi_x \end{cases}. \tag{2.24}
$$

Here, leading and lagging HF legs are identified as $j = 1$ and 2, respectively. $i_{x,j}\big|_{P_{xy}}$ represents the jth leg turn-on current of port-x corresponding to the DAB cell comprising port-x and y bridge voltage sources (v_x and v_y) and the Δ-model inter-port inductor L_{xy} connected in between and the power is flowing from port-x to port-y ($\varphi_y > \varphi_x$). On the other hand, $i_{x,j}\big|_{P_{yx}}$ showcases the same turn-on current of that DAB cell if the power flow is in the reverse direction ($\varphi_x > \varphi_y$). Both of these current quantities can be derived from their expressions given in Table 2.2 based on the relative values of phase-duty control variables. Thus, the peak switching currents in a MAB converter can be calculated from the developed time-domain model of the converter fundamentally constructed using multiple DAB building blocks. These identified current vertices are major point of interest while computing the switching loss in the system as well as evaluating soft-switching ability of the converter. Furthermore, it is noteworthy to mention that, regardless of the power flow directions in a Δ-MAB model, if the fundamental DAB cells operate in ZVS-favorable modes (mode-1, 4 or 5), the superposition of the port currents creates a favorable condition for the MAB port switches to undergo soft-switching.

Furthermore, using the same modeling technique, the average normalized power sourced by any of the MAB port (port-x) can be precisely quantified using (2.23) and the expressions given in Table 2.2.

2.3   All Harmonics Inclusive Frequency Domain Modeling of M2PC

As observed from Sect. 2.2, the time domain-based modeling of any M2PC requires application of superposition theorem and mode dependent analysis of the fundamental DAB building block based on the relative values of the operating control variables. This process becomes computationally heavy as the number of MAB ports increases and additionally, unified expressions for power flow and winding current quantities cannot be derived. Further, although computing the switching instant winding currents for any H-bridge is possible using the time-domain approach, calculating the winding current RMSs in an entire switching cycle for a TAB and higher order MAB converter becomes significantly difficult due to increasing number of peaks in the winding currents appearing due to multiple switching instants. However, the transformer winding RMS current computation is a necessary step in modeling the conduction loss in such converters. In order to circumvent such challenges, and to provide unified expressions for the average power transfer, the instantaneous and RMS winding currents, a frequency domain analysis of DAB, TAB and n-port MAB converter is carried out and presented below.

2.3.1   Frequency Domain Modeling of DAB

The switching equivalent circuit of the DAB converter was presented earlier in Fig. 1.3, where the voltage sources v_1 and v_2, represent the quasi-square shaped bridge voltages of the primary and secondary H-bridges, respectively. The shapes of the v_1 and v_2, are defined by the phase shift (φ) between them and their respective duty cycles δ_1 and δ_2, as shown in Fig. 2.2. Applying Fourier series expansion on these voltages we get,

$$v_1(t) = \frac{4V_1}{\pi} \sum_{k=1}^{2m+1} \frac{1}{k} \cos(k\delta_1)\sin(k\omega_s t); \tag{2.25}$$

$$v_2(t) = \frac{4V_2}{\pi} \sum_{k=1}^{2m+1} \frac{1}{k} \cos(k\delta_2)\sin\{k(\omega_s t - \varphi)\}. \tag{2.26}$$

Where 'k' ($k = 1, 3, 5, \ldots, 2m + 1$; m is $+ve$ $integer$) is the order of the harmonic and $\omega_s = 2\pi f_{sw}, f_{sw}$ is the switching frequency of the converter. It is observed from (2.25) and (2.26) that, v_1 and v_2 can be expressed as a superposition of multiple sinusoidal voltage sources having the fundamental and its higher order odd harmonic frequencies. Thus, the equivalent circuit of the DAB converter can be redrawn as Fig. 2.13, which is called generalized harmonic approximation (GHA) based equivalent circuit model of the DAB converter. The reconstruction of the quasi-square shaped bridge voltages becomes more accurate as a higher number of harmonics (m) are kept under consideration. If only the first harmonic is considered in such frequency domain model, it is called a first-harmonic approximation (FHA). The reconstruction of the DAB switching bridge voltages with varying harmonic orders under consideration for a specific converter operating condition is presented in Fig. 2.14. The parameters of the DAB converter under this case study are outlined in Table 2.3.

In order to determine the conduction as well as the switching losses in a DAB converter, the instantaneous transformer winding currents and their RMSs must be accurately quantified as functions of the control variables (δ_1, δ_2 and φ). The inductor current (i_L) in a DAB can be calculated by applying the volt-sec balance law across the inductor L, and can be written as,

Fig. 2.13 GHA based modeling of the quasi-square wave shaped bridge voltages, approximated up to kth order

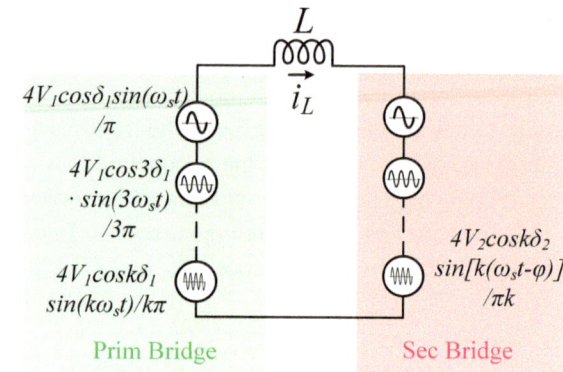

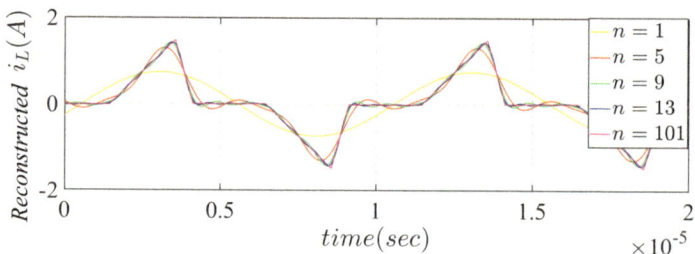

Fig. 2.14 Approximated inductor current shape considering different orders of harmonics using GHA model for $(\delta_1, \delta_2, \varphi) = (0.9065, 0.729, 0.176)$ under the test conditions defined as: $V_1 = 160V$, $V_2' = 90V$, and $P_o = 50W$

Table 2.3 Design specifications of the DAB converter under case study

Circuit parameters	Values
Input DC voltage (V_1)	160 V
Output DC voltage (V_2')	90–150 V, 120 V (nominal)
Output rated power (P_{out})	400 W
Transformer turns ratio (N$_1$:N$_2$)	7:5
Magnetizing inductance (L_m)	268 µH
Total leakage inductances (L_1, L_2')	16 and 15 µH
Output capacitor (C_{o2})	86 µF
Switching frequency (f_{sw})	100 kHz

$$i_L = \int \frac{v_L(t)dt}{L}, \tag{2.27}$$

where v_L is the voltage across the inductor L. Now, using the relations given in (2.25) and (2.26), v_L can be expressed as follows:

$$v_L(t) = v_1(t) - v_2(t)$$

$$= \frac{4}{\pi} \left[\sum_{k=1}^{2m+1} \frac{V_1}{k} \cos(k\delta_1)\sin(k\omega_s t) - \frac{V_2}{k} \cos(k\delta_2)\sin\{k(\omega_s t - \varphi)\} \right]. \tag{2.28}$$

Applying (2.28) in (2.27), the time varying inductor current is derived and simplified as (2.29).

$$i_L = \sum_{k=1}^{2m+1} [A_k \sin(k\omega_s t) + B_k \cos(k\omega_s t)] \tag{2.29}$$

where the harmonic order 'k' dependent coefficients are, $A_k = \frac{4}{\pi \omega L k^2}[V_2\cos(k\delta_2)\sin(k\varphi)]$ and $B_k = \frac{4}{\pi \omega L k^2}[V_2\cos(k\delta_2)\cos(k\varphi) - V_1\cos(k\delta_1)]$. Using (2.29), i_L can be determined during any time instant including the switching device turn-on and off instants.

Furthermore, in the GHA-based DAB converter network being analyzed, power transfer via the inductor L is feasible solely between two distinct sinusoidal voltage sources operating at the same frequency. By applying the superposition theorem and considering the harmonics up to the (2 m + 1)-th order, the power flow from the input to the output can be expressed as,

$$P_{12} = \sum_{k=1}^{2m+1} \left[\frac{v_{1,k,RMS} \times v_{2,k,RMS}}{X_L} \times \sin(k\varphi) \right]$$

$$= \sum_{k=1}^{2m+1} \left[\frac{4V_1\cos(k\delta_1)}{\pi k \sqrt{2}} \cdot \frac{4V_2\cos(k\delta_2)}{\pi k \sqrt{2}} \cdot \sin(k\varphi) \cdot \frac{1}{2\pi k f_{sw}L} \right]$$

$$= \frac{4V_1 V_2}{\pi^3 f_{sw} L} \cdot \sum_{k=1}^{2m+1} \frac{1}{k^3}[\cos(k\delta_2)\cos(k\delta_1)\sin(k\varphi)]. \tag{2.30}$$

This expression depends on the phase displacement between the H-bridge voltages (φ), their respective duty cycles (δ_1 and δ_2), switching frequency (f_{sw}) of the MOSFETs, the terminal DC voltages, and the leakage inductances of the transformer windings. To achieve a specified load power at particular output voltage levels, (2.30) can be solved for the control variables δ_1, δ_2 and φ, under a three variable controlled DAB modulation. It is also observed from the power flow expression that the direction of the power transfer only depends on φ. In other words, primary will transfer power to secondary if the secondary H-bridge voltage is lagging the primary bridge voltage, otherwise if v_2 is leading v_1, the power flow direction becomes reverse. Notably, this specifically holds true for a purely inductive tank-based converter of the MAB family.

Thus, using the higher harmonic inclusive unified expressions of the inductor current and the power transfer given in (2.29) and (2.30), the DAB converter can be easily and precisely modelled while avoiding the time-domain based segmentized formulation described in Sect. 2.2.1.

2.3.2 Frequency Domain Modeling of TAB

Similar to the frequency-domain oriented DAB modeling, the GHA based modeling approach can be applied to the TAB equivalent circuits presented in Fig. 2.7. The port or full-bridge voltages of quasi-square wave shapes can be decomposed in an infinite sum Fourier series. However, approximating the generalized harmonic model up to nth harmonic order, the ac port voltages can be presented as,

$$
\left.
\begin{aligned}
v_1(t) &= \frac{4V_1}{\pi} \sum_{k=1}^{n=2m+1} \frac{1}{k} \cos(k\delta_1) \sin(k\omega_s t) \\
v_2(t) &= \frac{4V_2}{\pi} \sum_{k=1}^{n=2m+1} \frac{1}{k} \cos(k\delta_2) \sin\{k(\omega_s t - \varphi_2)\} \\
v_3(t) &= \frac{4V_3}{\pi} \sum_{k=1}^{n=2m+1} \frac{1}{k} \cos(k\delta_3) \sin\{k(\omega_s t - \varphi_3)\}
\end{aligned}
\right\}
\tag{2.31}
$$

where 'k' is the order of the harmonic, m is a +ve integer, and $\omega_s = 2\pi f_{sw}$, f_{sw} being the switching frequency of the converter. Thus, each port voltage can be expressed as a series combination of sinusoidal voltage sources of $kf_{sw}(k = 1, 3, 5, \ldots, 2m + 1)$ frequency, as shown in the Δ-TAB circuit of Fig. 2.15.

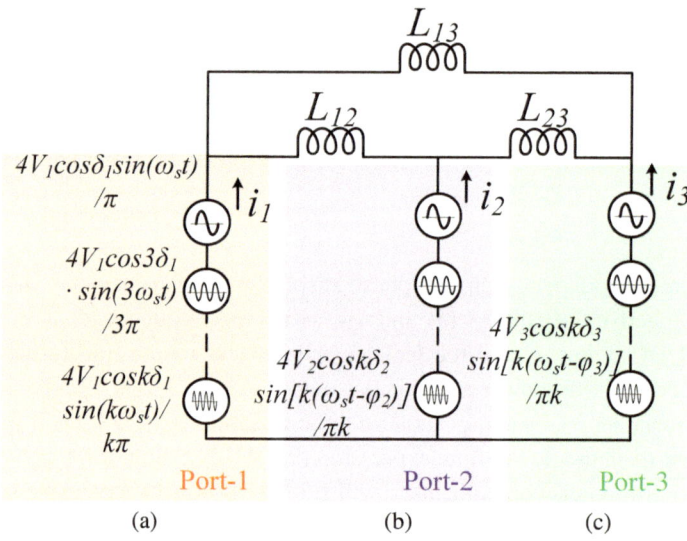

Fig. 2.15 **a** Y-equivalent and **b** Δ-equivalent circuit of TAB; **c** GHA based modeling of the port voltages, approximated up to 'k'th order

The line inductor current, flowing between two ports port-a and port-b, in the circuit described in Δ-convention can be calculated as, $i_{ab}(t) = \int_0^t \frac{v_{ab}(t)dt}{L_{ab}} = \frac{1}{L_{ab}}\int_0^t [v_a(t) - v_b(t)]dt$, where, i_{ab} is the current flowing from port-a to port-b and v_{ab} is the voltage across the inductance L_{ab}. Now, using the relations given in (2.31), the inter-port inductor currents in the Δ-equivalent circuit can be written as,

$$
\left.
\begin{aligned}
i_{12}(t) &= \frac{4}{\pi\omega_s L_{12}}\sum_{k=1}^{2m+1}\left[-\frac{V_1}{k^2}\cos(k\delta_1)\cos(k\omega_s t) + \frac{V_2}{k^2}\cos(k\delta_2)\cos\{k(\omega_s t - \varphi_2)\}\right]\\
i_{13}(t) &= \frac{4}{\pi\omega_s L_{13}}\sum_{k=1}^{2m+1}\left[-\frac{V_1}{k^2}\cos(k\delta_1)\cos(k\omega_s t) + \frac{V_3}{k^2}\cos(k\delta_3)\cos\{k(\omega_s t - \varphi_3)\}\right]\\
i_{23}(t) &= \frac{4}{\pi\omega_s L_{23}}\sum_{k=1}^{2m+1}\left[-\frac{V_2}{k^2}\cos(k\delta_2)\cos\{k(\omega_s t - \varphi_2)\} + \frac{V_3}{k^2}\cos(k\delta_3)\cos\{k(\omega_s t - \varphi_3)\}\right]
\end{aligned}
\right\}
$$
$$(2.32)$$

Applying KCL at the individual port nodes, the total current sourced by a port can be formulated such as:

$$
i_1(t) = i_{12}(t) + i_{13}(t). \tag{2.33}
$$

Using relations mentioned in (2.32) and (2.33), i_1 can be simplified as (2.34).

$$
i_1 = \sum_{k=1}^{2m+1}\left[A_{k,1}\sin(k\omega_s t) + B_{k,1}\cos(k\omega_s t)\right] \tag{2.34}
$$

where the harmonic order 'k' dependent coefficients are $A_{k,1} = \frac{4V_1}{\pi\omega k^2}\left[\frac{m_2\cos(k\delta_2)\sin(k\varphi_2)}{L_{12}} + \frac{m_3\cos(k\delta_3)\sin(k\varphi_3)}{L_{13}}\right]$ and $B_{k,1} = \frac{4V_1}{\pi\omega k^2}\left[\frac{m_2\cos(k\delta_2)\cos(k\varphi_2)-\cos(k\delta_1)}{L_{12}} + \frac{m_3\cos(k\delta_3)\cos(k\varphi_3)-\cos(k\delta_1)}{L_{13}}\right]$. The individual port's dc voltage gains with respect to port 1 are denoted as $m_2 = \frac{V_2}{V_1}$ and $m_3 = \frac{V_3}{V_1}$. Similarly, the instantaneous current expressions for i_1 and i_2 can be also derived as,

$$
i_2(t) = \sum_{k=1}^{2m+1}\left[A_{k,2}\sin(k\omega_s t) + B_{k,2}\cos(k\omega_s t)\right] \tag{2.35}
$$

$$
i_3(t) = \sum_{k=1}^{2m+1}\left[A_{k,3}\sin(k\omega_s t) + B_{k,3}\cos(k\omega_s t)\right] \tag{2.36}
$$

As indicated in Eqs. (2.34)–(2.36), the accuracy of the GHA based predicted wave-shape of instantaneous port currents is highly reliant on incorporating various harmonic components into the current formulation equations. To illustrate this point, Fig. 2.16 shows the winding current waveshapes for a specific set of operating control variables defined as $(\delta_1, \delta_2, \delta_3, \varphi_2, \varphi_3) = (0.12, 0.23, 0.16, 0.56, 0.32)$, which correspond to the design specifications outlined in Table 2.4. Furthermore, Table 2.5 provides a detailed comparison between the theoretically calculated switch turn-on current levels for switch S1 and the actual turn-on current observed from simulation-based analysis across different harmonic orders. It is evident that the theoretical values derived using (2.34) closely align with the simulation results for harmonic orders $n \geq 13$, with an error margin of less than 5%. Therefore, the proposed GHA model-based method for quantifying instantaneous port currents effectively bypasses the need for the traditional, complex time-domain analysis for winding current calculations.

Further, the power transfer in a TAB circuit can be modeled using the Δ-circuit and the already derived power transfer equation of the DAB circuit, given by (2.30). As we know power transfer can occur between any two sinusoidal voltage sources operating at the same frequency and this power exchange is influenced by the amplitudes of the ac bridge voltages, their phase difference, and the impedance of the connecting line. By applying the superposition theorem, the total power transferred from port-1 to port-2 can

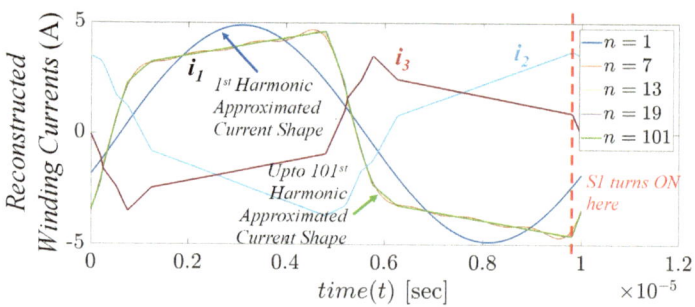

Fig. 2.16 Approximated i_1 current shape considering different orders of harmonics using GHA model for $(\delta_1, \delta_2, \delta_3, \varphi_2, \varphi_3, m_2, m_3) = (0.12, 0.23, 0.16, 0.56, 0.32, 0.8, 1)$

Table 2.4 TAB circuit parameters

Circuit parameters	Values
Input DC voltage (V_1)	160 V
Secondary output port voltage (V_2)	100–140 V, 120 V (nominal)
Secondary output port rated power (P_2)	400 W
Tertiary output port voltage (V_3)	16–28 V, 22 V (nominal)
Tertiary output port rated power (P_3)	400 W
Transformer turns ratio (N_1, N_2, N_3)	7:5:1
Magnetizing inductance (L_m)	300 μH
Total leakage inductances (L_1, L_2', L_3')	16, 15 and 0.28 μH
Secondary side output capacitor (C_2)	86 μF
Tertiary output capacitor (C_3)	47 μF
Switching frequency (f_{sw})	100 kHz

Table 2.5 Comparison between GHA predicted and simulated $i_{S1,\,ON}$

| Control variables | Considered | Calculated $i_{S1,\,ON}$ (A) | | Error in theoretical |
$(\delta_1, \delta_2, \delta_3, \varphi_2, \varphi_3,$ $m_2, m_3)$	harmonic order (n)	GHA model based	Simulation	calculation (%)
(0.12, 0.23, 0.16,	1	− 2.3508	− 4.5994	48.889
0.56, 0.32, 0.8, 1)	7	− 4.3008		6.492
	13	− 4.3896		4.561
	19	− 4.4424		3.413

be expressed as:

$$P_{12} = \frac{4}{\pi^3 f_{sw}} \cdot \sum_{k=1}^{2m+1} \frac{1}{k^3} \left[\frac{V_1 V_2}{L_{12}} \cos(k\delta_1) \cos(k\delta_2) \sin\{k(\varphi_2 - \varphi_1)\} \right]. \tag{2.37}$$

This power flow can be normalized with respect to port-1 as,

$$P_{12} = \frac{8P_{base}}{\pi^2} \cdot \sum_{k=1}^{2m+1} \frac{1}{k^3} \left[\frac{3m_2 L_1}{L_{12}} d_{k1} d_{k2} \sin\{k(\varphi_2 - \varphi_1)\} \right], \tag{2.38}$$

where $P_{base} = \frac{V_1^2}{2\pi f_{sw} L_1}$, $d_{k1} = \cos(k\delta_1)$ and $d_{k2} = \cos(k\delta_2)$.

Considering port-2 and port-3 as output ports and port-1 as the input port in this three-port system, the total power absorbed at port-2 and port-3 is represented in (2.39) and (2.40), respectively.

$$P_2 = P_{12} + P_{32} = \frac{8P_{base}}{\pi^2} \sum_{k=1}^{2m+1} \frac{1}{k^3} \left[\frac{3m_2 L_1}{L_{12}} d_{k1} d_{k2} \sin(k\varphi_2) \right.$$
$$\left. + \frac{3m_2 m_3 L_1}{L_{23}} d_{k2} d_{k3} \sin\{k(\varphi_2 - \varphi_3)\} \right]. \tag{2.39}$$

$$P_3 = P_{13} + P_{23} = \frac{8P_{base}}{\pi^2} \sum_{k=1}^{2m+1} \frac{1}{k^3} \left[\frac{3m_3 L_1}{L_{13}} d_{k1} d_{k3} \sin(k\varphi_3) \right.$$
$$\left. + \frac{3m_2 m_3 L_1}{L_{23}} d_{k2} d_{k3} \sin\{k(\varphi_3 - \varphi_2)\} \right]. \tag{2.40}$$

Due to the interconnected nature of the TAB architecture, the power delivered to any port is observed to be influenced by the phase shifts of either of the output ports.

2.3.3 Generalized Frequency Domain Modeling of n-Port MAB

Following the frequency-domain circuit modeling of the DAB and TAB dc-dc converter, a GHA based frequency domain analysis of a generalized n-port MAB converrter is carried out and presented in this section. The quasi-square shaped port voltages (v_i) of any MAB port from its Y- and Δ-equivalent circuits can be mathematically expanded in infinite Fourier series and thus represented as a series combination of sinusoidal voltage sources with odd order harmonics, depicted in Fig. 2.17.

The ac bridge voltage of the ith port can be represented as:

$$v_i'(t) = \frac{4V_i'}{\pi} \cdot \sum_{k=1,3,\ldots}^{\infty} \frac{1}{k} d_{ki} \sin\{k(\omega_s t - \varphi_i)\} \tag{2.41}$$

where $d_{ki} = \cos(k\delta_i)$; 'k' is the order of the harmonic, and $\omega_s = 2\pi f_{sw}$, f_{sw} being the switching frequency of the converter.

From the Δ-model of MAB, the current flowing from port i to port j through their corresponding link inductor L_{ij} can be deduced as,

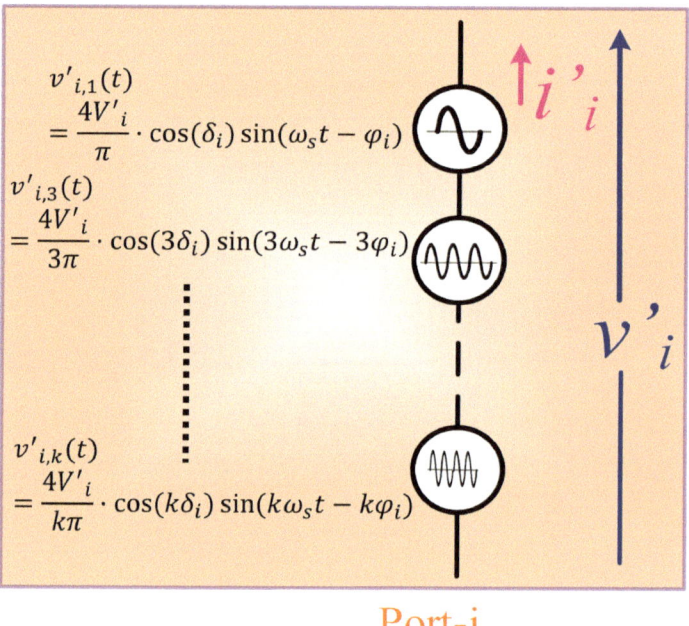

Port-i

Fig. 2.17 GHA model based sinusoidal approximation of quasi-square shaped any AB voltage, $v_i{'}$

$$i_{ij}(t) = \frac{1}{L_{ij}} \int_0^t \left[v_i(t) - v_j(t) \right] dt$$

$$= \frac{4}{\pi \omega_s L_{ij}} \cdot \sum_{k=1,3,\ldots}^{\infty} \left[-\frac{V_i}{k^2} d_{ki} \cos\{k(\omega_s t - \varphi_i)\} + \frac{V_j}{k^2} d_{kj} \cos\{k(\omega_s t - \varphi_j)\} \right]. \qquad (2.42)$$

Applying KCL at the individual port nodes, the generalized expression of the total current sourced by any MAB port-i can be formulated and simplified as (2.43).

$$i_i(t) = \sum_{k=1,3,5,\ldots}^{\infty} \left(A_{i,k} \cos(k\omega_s t) + B_{i,k} \sin(k\omega_s t) \right). \qquad (2.43)$$

where the harmonic order dependent coefficients $A_{i,k}$ and $B_{i,k}$ are:

$$A_{i,k} = \frac{4}{\pi \omega_s k^2} \left[\sum_{\substack{j=1 \\ j \neq i}}^{n} \frac{1}{L_{ij}} \{ -V_i d_{ki} \cos(k\varphi_i) + V_j d_{kj} \cos(k\varphi_j) \} \right]; \qquad (2.44)$$

$$B_{i,k} = \frac{4}{\pi \omega_s k^2} \left[\sum_{\substack{j=1 \\ j \neq i}}^{n} \frac{1}{L_{ij}} \{ -V_i d_{ki} \sin(k\varphi_i) + V_j d_{kj} \sin(k\varphi_j) \} \right]. \qquad (2.45)$$

The already known expression for the power flow between two of the MAB ports (port-i and j) using the GHA method is written in (2.46).

$$P_{ij} = \frac{4 V_i V_j}{\pi^3 f_{sw} L_{ij}} \cdot \sum_{k=1,3,\ldots}^{\infty} \frac{1}{k^3} \left[d_{ki} d_{kj} \sin\{k(\varphi_j - \varphi_i)\} \right] \qquad (2.46)$$

It can be inferred from (2.46) that the magnitude of the power transfer between port-i and port-j depends on their corresponding ac voltage port's phase difference $(\varphi_j - \varphi_i)$, their respective duty cycles (d_{ki} and d_{kj}), dc link voltages (V_i and V_j) and the equivalent inter-port inductance (L_{ij}). The leading or lagging nature of port voltage phase difference, i.e., $(\varphi_j - \varphi_i)$ determines the direction of the power flow. The expression of the total power supplied by the any of the MAB port (ith port) can be quantified as the summation of the power transfers to the rest of the ports, which is represented in (2.47).

$$P_i = \sum_{\substack{j=1 \\ j \neq i}}^{n} P_{ij} = \sum_{\substack{j=1 \\ j \neq i}}^{n} \frac{4V_i V_j}{\pi^3 f_{sw} L_{ij}} \cdot \sum_{k=1,3,...}^{\infty} \frac{1}{k^3}\left[d_{ki}d_{kj}\sin\{k(\varphi_i - \varphi_j)\}\right] \tag{2.47}$$

2.4   RMS Current Modeling for Full-Range M2PC Operation

The utilization of the GHA model to derive a unified expression for the transformer winding current or bridge output current in an n-port Multi-Active Bridge (MAB) converter
simplifies the RMS calculation significantly compared to traditional time-domain modeling. By employing the continuous time current expression of i_i as given in (2.43), its
RMS value can be efficiently determined using (2.48).

$$i_{i,RMS}^2 = \frac{1}{2}\sum_{k=1,3,...}^{\infty} \frac{1}{k^4}\left[A_{i,k}^2 + B_{i,k}^2\right] \tag{2.48}$$

This expression can be directly applied to compute the winding current RMS values in
any MAB converter, which is particularly crucial when quantifying the conduction losses
in the MOSFETs and transformer windings. As noted in (2.42) and (2.48), the precision of the GHA-derived waveshape of both instantaneous and RMS port currents heavily
relies on the inclusion of the number of harmonic components in the current formulation equations. To illustrate this, Fig. 2.18 shows the port-1 winding current waveshape
of a Quadruple-Active-Bridge (QAB) converter for a set of operating control variables
defined as $(\delta_1, \delta_2, \delta_3, \delta_4, \varphi_2, \varphi_3, \varphi_4) = (0.6, 0, 0, 0.8, 0.2, 0.3, 0.4)$, according to the
design specifications mentioned in Table 2.6.

Additionally, Table 2.7 provides a detailed comparison between the theoretically
obtained port-1 winding RMS current and the switch turn-on current level for switch S11,

Fig. 2.18 Approximated i_1
current shape of a QAB
converter considering different
orders of harmonics using
GHA model, for
$(\delta_1, \delta_2, \delta_3, \delta_4, \varphi_2, \varphi_3, \varphi_4) =$
$(0.6, 0, 0, 0.8, 0.2, 0.3, 0.4)$

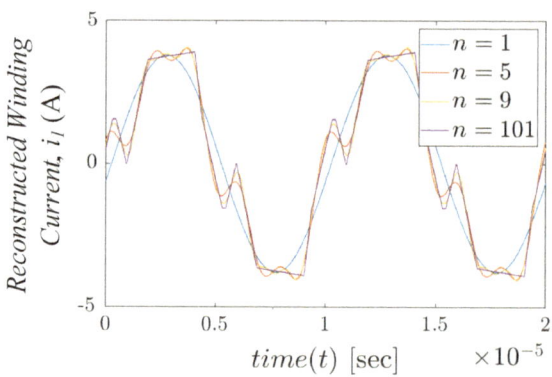

Table 2.6 Circuit parameters of QAB under study

Circuit parameters	Values
Port-1 DC Link (V_1)	160 V
Port-2 DC Link (V'_2)	90–150 V, 120 V (nominal)
Port-3 DC Link (V'_3)	90–150 V, 120 V (nominal)
Port-4 DC Link (V'_4)	16–28 V, 22 V (nominal)
Port-1 rated power (P_1)	400 W
Port-2 rated power (P_2)	200 W
Port-3 rated power (P_3)	200 W
Port-4 rated power (P_4)	200 W
Transformer turns ratio (n_1:n_2:n_3: n_4)	7:5:5:1
Magnetizing inductance (L_m)	360 μH
Leakage inductances (L'_1, L_2', L'_3, L_4')	16, 15.4, 15.5, and 0.27 μH
DC link capacitors (C_{o1}, C_{o2}, C_{o3}, C_{o4})	258, 258, 258, and 398 μF
Switching frequency (f_{sw})	100 kHz

evaluated for different harmonic orders. It is observed that the GHA model-oriented winding RMS current values obtained using (2.48) align closely with the simulation results for harmonic orders n ≥ 5, maintaining an error margin of less than 5%.

Furthermore, the data indicate that to accurately estimate the turn-on and turn-off current values, which are essential for estimating the switching loss of the QAB converter, a higher order switching harmonic needs to be included specifically, n ≥ 13 for an error margin of less than 5%.

Table 2.7 Comparison between GHA model predicted and simulated port-1 RMS current and S11 switch turn-on current

Control variables (δ_1, δ_2, δ_3, δ_4, φ_2, φ_3, φ_4)	Considered harmonic order (n)	Calculated $i'_{1, rms}$ (A)		Error in analytical model (%)	Calculated $i_{S1,ON}$ (A)		Error in analytical model (%)
		GHA model based	Simulation		GHA model based	Simulation	
(0.6, 0, 0, 0.8, 0.2, 0.3, 0.4)	1	2.34	2.6047	10.162	0.29	1.55	81.290
	3	2.42		7.091	0.86		44.516
	5	2.541		2.446	1.09		29.677
	7	2.572		1.255	1.42		8.387
	13	2.6		0.180	1.49		3.871

From this study, it is evident that there is a trade-off between the considered harmonic order and the accuracy of the estimated MAB port current RMS and instantaneous currents. Higher harmonic orders improve accuracy but may increase computational complexity for online calculation of RMS currents. This balance is crucial for optimizing the performance and efficiency of MAB converters, particularly in minimizing conduction and switching losses.

In summary, the GHA model-based approach provides a more straightforward and accurate method for RMS current calculation in MAB converters, facilitating better design and control strategies for these systems.

Utilizing (2.48), mathematically simplified expressions of the ac port currents of a TAB converter are derived in (2.49)–(2.51).

$$
\begin{aligned}
i_{1,rms}^2 &= \frac{1}{2} \sum_{k=1}^{2m+1} \frac{1}{k^4} \left[A_k^2 + B_k^2 \right] = \frac{8V_1^2}{\pi^2 \omega^2} \cdot \sum_{k=1}^{2m+1} \frac{1}{k^4} \left[d_{k1}^2 \left(\frac{1}{L_{12}^2} + \frac{1}{L_{13}^2} + \frac{2}{L_{12}L_{13}} \right) \right. \\
&\quad + \frac{m_2^2 d_2^2}{L_{12}^2} + \frac{m_3^2 d_3^2}{L_{13}^2} - 2 d_{k1} d_{k2} m_2 \cos(k\varphi_2) \left(\frac{1}{L_{12}^2} + \frac{1}{L_{12}L_{13}} \right) \\
&\quad - 2 d_{k1} d_{k3} m_3 \cos(k\varphi_3) \left(\frac{1}{L_{13}^2} + \frac{1}{L_{12}L_{13}} \right) \\
&\quad \left. + 2 m_2 m_3 d_{k2} d_{k3} \cos\{k(\varphi_2 - \varphi_3)\} \left(\frac{1}{L_{12}L_{13}} \right) \right]
\end{aligned}
\tag{2.49}
$$

$$
\begin{aligned}
i_{2,rms}^2 &= \frac{8V_1^2}{\pi^2 \omega^2} \cdot \sum_{k=1}^{2m+1} \frac{1}{k^4} \left[(m_2 d_{k2})^2 \left(\frac{1}{L_{12}^2} + \frac{1}{L_{23}^2} + \frac{2}{L_{12}L_{23}} \right) + \frac{d_{k1}^2}{L_{12}^2} + \frac{(m_3 d_{k3})^2}{L_{23}^2} \right. \\
&\quad - 2 d_{k1} d_{k2} m_2 \cos(k\varphi_2) \left(\frac{1}{L_{12}^2} + \frac{1}{L_{12}L_{23}} \right) \\
&\quad - 2 m_2 m_3 d_{k2} d_{k3} \cos\{k(\varphi_2 - \varphi_3)\} \left(\frac{1}{L_{23}^2} + \frac{1}{L_{12}L_{23}} \right) \\
&\quad \left. + 2 d_{k1} d_{k3} m_3 \cos(k\varphi_3) \left(\frac{1}{L_{12}L_{23}} \right) \right]
\end{aligned}
\tag{2.50}
$$

$$
\begin{aligned}
i_{3,rms}^2 &= \frac{8V_1^2}{\pi^2 \omega^2} \cdot \sum_{k=1}^{2m+1} \frac{1}{k^4} \left[(m_3 d_{k3})^2 \left(\frac{1}{L_{13}^2} + \frac{1}{L_{23}^2} + \frac{2}{L_{13}L_{23}} \right) + \frac{d_1^2}{L_{13}^2} + \frac{(m_2 d_{k2})^2}{L_{23}^2} \right. \\
&\quad - 2 d_{k1} d_{k3} m_3 \cos(k\varphi_3) \left(\frac{1}{L_{13}^2} + \frac{1}{L_{13}L_{23}} \right) \\
&\quad - 2 m_2 m_3 d_{k2} d_{k3} \cos\{k(\varphi_3 - \varphi_2)\} \left(\frac{1}{L_{23}^2} + \frac{1}{L_{13}L_{23}} \right)
\end{aligned}
$$

$$+2d_{k1}d_{k2}m_2\cos(k\varphi_2)\left(\frac{1}{L_{13}L_{23}}\right)\Bigg]$$ (2.51)

2.5 Modeling Soft-Switching Criterion in M2PC Switching Cells

The way the RMS values of the transformer winding currents or the H-bridge output currents of a M2PC plays direct role in defining the conduction losses in the system, the switching loss in such converters are dictated by the instantaneous value of the winding currents at switching instants of the H-bridge legs. Moreover, the direction and amplitude of such currents during the turn-on and turn-off instants decides if the MOSFETs will undergo soft turn-on (ZVS) and soft turn-off (zero-current switching or ZCS). Normally, in power MOSFETs utilizing traditional Si technology or wide-bandgap semiconductors such as SiC and GaN, the turn-on loss is significantly higher compared to the turn-off losses. Therefore, in order to attain better system efficiency and improved EMI performance for a wide load operation, the achievement of ZVS operation in the MAB converter is of paramount importance. To realize effective ZVS operation, it is crucial to consider the charge or energy stored in the non-linear parasitic capacitors (C_{OSS}) of the switching power devices during the commutation process. This necessitates a minimal inductor current ($I_{Lx,crit}$) to successfully complete the commutation process during the dead-band, a parameter that will be comprehensively derived in this section. Before diving into deriving the conditions for ZVS in a n-port MAB converter, a generalized Thevenin equivalent circuit model of the converter's commutating H-bridge is derived so that a unified condition of soft-switching can be formulated that is applicable for any converter under the MAB family consisting of any number of H-bridge cells within its architecture.

2.5.1 Thevenin Equivalent Circuit Formulation

The equivalent circuit related to Port-x (where x signifies the commutating port) is depicted in Fig. 2.19. In this representation, V_x represents the dc link voltage on the commutating side, while v_x and i_{Lx} denote the voltage across the half-bridges and the port-x inductor current on the commutating side, respectively. $L_{TH,x}$ and $v_{TH,x}$ represent the Thevenin equivalent impedance and voltage of the non-commutating side. The expressions for $L_{TH,x}$ and $v_{TH,x}$ are derived from the Y-type MAB configuration based on the Thevenin theorem as shown in Fig. 2.20 and are presented in (2.52) and (2.53).

$$L_{TH,x} = \left(\sum_{\substack{y=1\\y\neq x}}^{n}\frac{1}{L_y}\right)^{-1} + L_x.$$ (2.52)

$$v_{TH,x}(t) = \left(\sum_{\substack{y=1 \\ y \neq x}}^{n} \frac{v_y(t)}{L_y} \right) \cdot \left(\sum_{\substack{y=1 \\ y \neq x}}^{n} \frac{1}{L_y} \right)^{-1}. \tag{2.53}$$

Here, $v_{TH,x}(t)$ contains the information regarding phase-duty control parameters of the non-commutating MAB ports and can be utilized in order to find the ZVS conditions of the port-x MOSFETs. At any switching time instant, $v_{TH,x}(t)$ can be derived from MAB port voltages $v_y(t)$ as given in (2.54).

Fig. 2.19 Thevenin equivalent circuit referred to Port-x of the TAB converter with equivalent line inductance $L_{th,x}$ and equivalent port voltage v_x

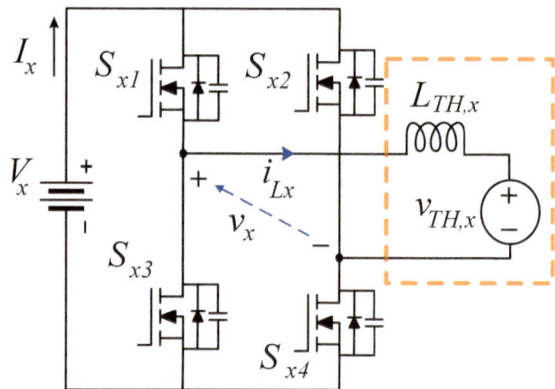

Fig. 2.20 Steps to derive Thevenin-equivalent circuit elements from MAB Y-model

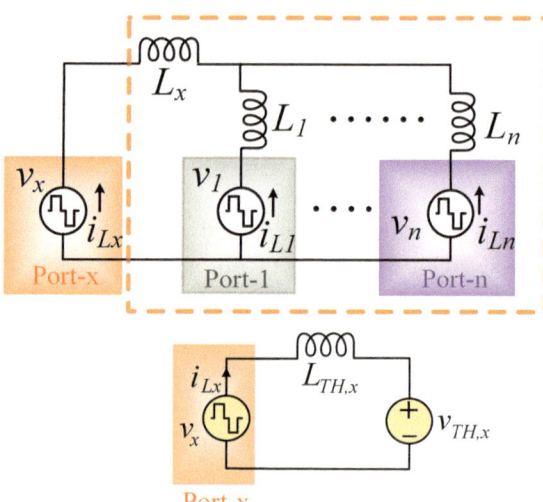

$$v_y(t) = \begin{cases} 0, & (\varphi_y - \delta_y) < \omega_s t < (\varphi_y + \delta_y) \\ V_y, & (\varphi_y + \delta_y) < \omega_s t < (\pi + \varphi_y - \delta_y) \\ 0, & (\pi + \varphi_y - \delta_y) < \omega_s t < (\pi + \varphi_y + \delta_y) \\ -V_y, & (\pi + \varphi_y + \delta_y) < \omega_s t < (2\pi + \varphi_y - \delta_y). \end{cases} \tag{2.54}$$

Once the port equivalent circuit of the generalized MAB converter is synthesized, the ZVS conditions for the switching legs of that H-bridge can be identified by analyzing different circuit configurations that will appear based on the switching commutation instants. In the literature, the ZVS conditions for any MOSFET connected in a switching leg are analyzed using two different approaches: the energy-based ZVS and charge-based ZVS. Both of these approaches are explored to identify the ZVS conditions in a MAB converter and are mentioned in next sections.

2.5.2   Solutions for Energy-Based ZVS Conditions

For a generic MAB full-bridge (port-x), the equivalent circuit is presented in Fig. 2.19. For this configuration, six separate possible switching modes and their subsequent ZVS conditions are identified that encompass all unique commutations, as illustrated in Fig. 2.21. Any other switching mode will lead to hard-switching conditions. The six distinctly identified commutation cases can be divided into two groups: 1 and 2, based on the presence of duty cycle phase shift (δ_x). The ZVS conditions corresponding to each of the operating modes are synthesized and described below.

A. **Case 1(a)**: $\delta_x = 0 \, \& \, i_{Lx}(t) > 0$, S_{x2} and S_{x3} turning on

In this case of switching transition, S_{x1} and S_{x4} are turned-off while S_{x2} and S_{x3} are going to turn-on. Thus, during the dead time period, the inductor current i_{Lx} is going to charge the body capacitors ($C_{OSS,x}$) of $S_{x,1}$ and $S_{x,4}$ from 0 V to V_x while the body capacitors of S_{x2} and S_{x3} will discharge from V_x to 0V. Capacitance of the body capacitor of the MOSFET varies with applied drain-source voltage; thus, it can be expressed as a function of v_{ds}, i.e., $C_{OSS,x}(v_{ds})$. The nature of $C_{OSS,x}(v_{ds})$ with varying v_{ds} can be identified from the datasheet of the MOSFET.

As Fig. 2.21 suggests, in this circuit configuration, the total energy sunk by the sources is,

$$E_{sunk} = \int_0^{\tau_c} \left[v_{TH,x}(\tau) i_{Lx}(\tau) - V_x I_x(\tau) \right] dt \tag{2.55}$$

where $i_{Lx}(\tau)$ can be represented as $2C_{OSS,x}(v_{ds,S_{x1}}) \frac{dv_{ds,S_{x1}}}{dt}$. Further, as no portion of the inductor current leaks to the dc voltage source V_x and only circulates within the H-bridge circuit, $I_x(\tau) = 0$. Hence, (2.55) boils down to:

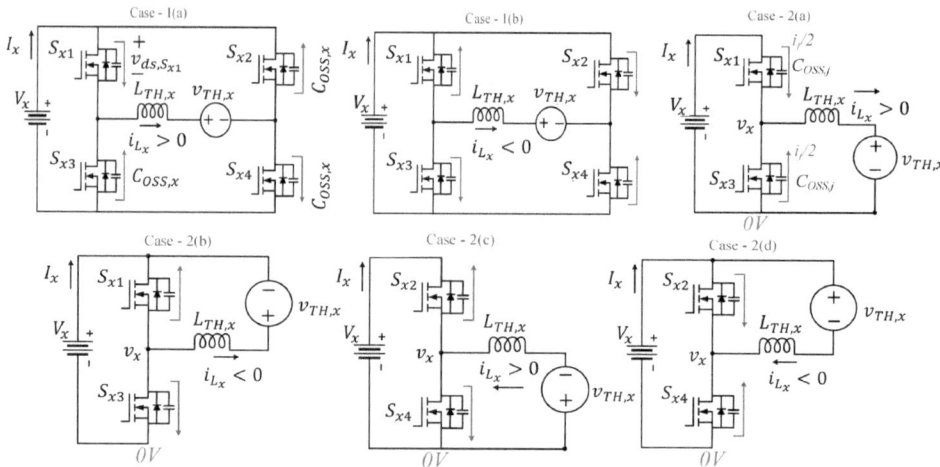

Fig. 2.21 Equivalent circuits for ZVS criteria study for different switching transient events: Case 1(a) to Case 2(d)

$$E_{sunk} = \int_0^{\tau_c} \left[v_{TH,x}(\tau) \left(2C_{OSS,x}(v_{ds,S_{x1}}) \frac{dv_{ds,S_{x1}}}{dt} \right) - 0 \right] dt$$

$$= \int_0^{V_x} v_{TH,x}(\tau) \left(2C_{OSS,x}(v_{ds,S_{x1}}) dv_{ds,S_{x1}} \right) = 2C_{Q,x}(V_x) v_{th,x}(\tau) V_x \qquad (2.56)$$

Here, $C_{Q,x}(V_x)$ is the charge-equivalent body capacitance of the MOSFET S_{x1} for V_x voltage applied across its drain-source.

Moreover, as the total stored energy in the MOSFET body capacitors does not change during the commutation interval, the ZVS will occur if the inductor has more stored energy compared to the energy sunk by the source. Thus, the condition for ZVS is:

$$\frac{1}{2} L_{TH,x} i_{Lx}(\tau)^2 \geq 2C_{Q,x}(V_x) v_{TH,x}(\tau) V_x. \qquad (2.57)$$

The relation of (2.57) suggests that for $v_{TH,x}(\tau) < 0$, ZVS will always be achieved. However, if $v_{TH,x}(\tau) V_x$ is positive, the minimum required port output current to attain ZVS can be expressed as,

$$|i_{Lx}(\tau)| \geq 2V_x \sqrt{\frac{C_{Q,x}(V_x)}{L_{TH,x}} \frac{v_{TH,x}(\tau)}{V_x}} \qquad (2.58)$$

where, τ is the turn on time instant of $S_{x,2}$, i.e., $(\varphi_x + \pi)$ and $|i_{Lx}(\tau)|$ can be attained from the winding current expression given in (2.43) while putting $\omega_s t = \varphi_x + \pi$.

B. **Case 1(b)**: $\delta_x = 0$ & $i_{Lx}(t) < 0$, S_{x1} and S_{x4} *turning on*

This switching case depicted in Fig. 2.21 represents the circuit condition during the dead time interval where S_{x2}, S_{x3} are turning-off and S_{x1}, S_{x4} are going to turn-on. The winding current, $i_{Lx}(\tau)$, is flowing into the full bridge, opposite to case 1(a). Forming the circuit equations in similar fashion as in case 1(a), the condition for ZVS is obtained as:

$$\frac{1}{2}L_{TH,x}i_{Lx}(\tau)^2 \geq -2C_{Q,x}(V_x)v_{TH,x}(\tau)V_x \tag{2.59}$$

The above relation implies that the minimal required inductor current for ZVS is (for $v_{TH,x}(\tau) < 0$)

$$|i_{Lx}(\tau)| \geq 2V_x\sqrt{-\frac{C_{Q,x}(V_x)}{L_{TH,x}}\frac{v_{TH,x}(\tau)}{V_x}} \tag{2.60}$$

where, τ is to be taken as the turn on instant of S_{x1} or S_{x4}, i.e., φ_x.

Observing the switching instants (τ) of Case 1(a) and 1(b), it is noticed that both the ZVS conditions given by (2.58) and (2.60) are actually the same (due to $v_{TH,x}(\varphi_x + \pi) = -v_{TH,x}(\varphi_x)$). In other words, for only phase shift-based control for any full bridge cell (i.e., $\delta_x = 0$), the viability of ZVS for all four switches can be concluded by checking any one condition between (2.58) and (2.60).

C. **Case 2(a)**: $\delta_x \neq 0$ & $i_{Lx}(t) > 0$, S_{x3} *turns-on*

The inclusion of duty cycle control for any H-bridge of the MAB introduces four switching cases (Case 2(a) to 2(d)) where the ZVS conditions for the two switching legs of a full bridge are different. For case 2(a), as depicted in Fig. 2.21, where the low side switch of the leading leg is turning on, the total energy absorbed by the sources can be expressed as,

$$E_{sunk} = \int_0^{\tau_c}\left[v_{TH,x}i_{Lx}(\tau) - V_xI_x(\tau)\right]dt$$

$$= \int_0^{\tau_c}\left[v_{TH,x}\left(2C_{OSS,x}(v_{ds,S_{x1}})\frac{dv_{ds,S_{x1}}}{dt}\right) - V_x\left(C_{OSS,x}(v_{ds,S_{x1}})\frac{dv_{ds,S_{x1}}}{dt}\right)\right]dt$$

$$= \int_0^{V_x}\left[2C_{OSS,x}(v_{ds,S_{x1}})v_{TH,x} - V_xC_{OSS,x}(v_{ds,S_{x1}})\right]dv_{ds,S_{x1}}$$

$$= 2C_{Q,x}(V_x)v_{TH,x}(\tau)V_x - C_{Q,x}(V_x)V_x^2 \tag{2.61}$$

Now, the minimum inductor current requirement for ZVS of $S_{x,2}$ can be obtained from,

$$\frac{1}{2}L_{TH,x}i_{Lx}(\tau)^2 \geq 2C_{Q,x}(V_x)v_{TH,x}(\tau)V_x - C_{Q,x}(V_x)V_x^2 \tag{2.62}$$

where, τ denotes the timing of S_{x3} turn-on, i.e., $(\varphi_x + \pi + \delta_x)$.

D. **Case 2(b)**: $\delta_x \neq 0$ & $i_{Lx}(t) < 0$, S_{x1} *turns-on*

Under this scenario, the high side switch of the leading leg turns on. The equivalent circuit is shown in Fig. 2.21.

Similar to previous cases, the total energy sunk by the sources is determined:

$$E_{sunk} = \int_0^{\tau_c} \left[v_{TH,x}(\tau)i_{Lx}(\tau) - V_xI_x(\tau)\right]dt$$

$$= \int_0^{\tau_c} \left[v_{TH,x}(\tau)\left(2C_{OSS,x}\left(v_{ds,S_{x1}}\right)\frac{dv_{ds,S_{x1}}}{dt}\right) - V_x\left(C_{OSS,x}\left(v_{ds,S_{x1}}\right)\frac{dv_{ds,S_{x1}}}{dt}\right)\right]dt$$

$$= \int_{V_x}^0 \left[2C_{OSS,x}\left(v_{ds,S_{x1}}\right)v_{TH,x}(\tau) - V_xC_{OSS,x}\left(v_{ds,S_{x1}}\right)\right]dv_{ds,S_{x1}}$$

$$= -2C_{Q,x}(V_x)v_{TH,x}(\tau)V_x - C_{Q,x}(V_x)V_x^2 \tag{2.63}$$

For ZVS, $E_{sourced} > E_{sunk}$, i.e.,

$$\frac{1}{2}L_{TH,x}i_{Lx}(\tau)^2 \geq -2C_{Q,x}(V_x)v_{TH,x}(\tau)V_x - C_{Q,x}(V_x)V_x^2 \tag{2.64}$$

where, $\tau = (\varphi_x + \delta_x)$.

Considering the turn-on instants of case 2(a) and 2(b), the individual ZVS conditions given by (2.62) and (2.64) eventually indicate the same condition. Thus, it can be concluded both the switches connected in any half bridge of a TAB network experiences similar switching (hard or soft) phenomena depending on the winding current direction and value.

E. **Case 2(c)**: $\delta_x \neq 0$ & $i_{Lx}(t) > 0$, S_{x2} *turns-on*

After this switching transition, the high side switch of the lagging leg undergoes turn-on event. The ZVS conditions for this case can be identified as (2.65) and (2.66).

$$\frac{1}{2}L_{TH,x}i_{Lx}(\tau)^2 \geq 2C_{Q,x}(V_x)v_{TH,x}(\tau)V_x + C_{Q,x}(V_x)V_x^2 \tag{2.65}$$

$$|i_{Lx}(\tau)| \geq 2V_x\sqrt{\frac{C_{Q,x}(V_x)}{2L_{th,x}} + \frac{C_{Q,x}(V_x)}{L_{th,x}}\frac{v_{TH,x}(\tau)}{V_x}}, \text{ for } v_{TH,x}(\tau) > -\frac{V_x}{2} \tag{2.66}$$

F. **Case 2(d)**: $\delta_x \neq 0 \,\&\, i_{Lx}(t) < 0$, S_{x4} *turns-on*

The ZVS conditions for this low side turn on event of the lagging leg are also determined similarly and presented in (2.67) and (2.68).

$$\frac{1}{2}L_{TH,x}i_{Lx}(\tau)^2 \geq -2C_{Q,x}(V_x)v_{TH,x}(\tau)V_x + C_{Q,x}(V_x)V_x^2 \qquad (2.67)$$

$$|i_{Lx}(\tau)| \geq 2V_x\sqrt{\frac{C_{Q,x}(V_x)}{2L_{TH,x}} - \frac{C_{Q,x}(V_x)}{L_{TH,x}}\frac{v_{TH,x}(\tau)}{V_x}}, \text{ for } v_{TH,x}(\tau) > \frac{V_x}{2} \qquad (2.68)$$

The switching times for cases 2(c) and 2(d) indicate that Eqs. (2.66) and (2.68) represent the same conditions. Therefore, either equation can be used to verify the probability of ZVS for the lagging leg switches. The derived ZVS conditions for all possible switching scenarios of a generic TAB full-bridge cell are summarized in Table 2.8.

It can be observed that the energy-related ZVS conditions primarily depend on the charge-equivalent $C_{OSS,x}$ of the MOSFETs connected in the switching leg of an MAB H-bridge.

If the direction of the bridge current, i_{Lx}, aligns with the ZVS-favorable current direction as outlined in the switching commutation cases above, but the corresponding ZVS condition in Table 2.7 is not met, the switching-on transition of the switching leg will result in losses due to incomplete or partial soft-switching/ZVS. This scenario indicates that the inductor current is insufficient to fully discharge the $C_{OSS,x}$ of the MOSFET that

Table 2.8 Summarized ZVS conditions for MAB

Case	Switch undergoing turn-on event	Condition for hard-switching	Condition for complete ZVS	Switching time (τ)
1(a)	S_{x2} and S_{x3}	$i_{Lx}(\tau) < 0$	$\frac{1}{2}L_{TH,x}i_{Lx}(\tau)^2 \geq 2C_{Q,x}(V_x)v_{TH,x}(\tau)V_x$	$\varphi_x + \pi$
1(b)	S_{x1} and S_{x4}	$i_{Lx}(\tau) > 0$	$\frac{1}{2}L_{TH,x}i_{Lx}(\tau)^2 \geq -2C_{Q,x}(V_x)v_{TH,x}(\tau)V_x$	φ_x
2(a)	S_{x3}	$i_{Lx}(\tau) < 0$	$\frac{1}{2}L_{TH,x}i_{Lx}(\tau)^2 \geq$ $2C_{Q,x}(V_x)v_{TH,x}(\tau)V_x - C_{Q,x}(V_x)V_x^2$	$\varphi_x + \pi - \delta_x$
2(b)	S_{x1}	$i_{Lx}(\tau) > 0$	$\frac{1}{2}L_{TH,x}i_{Lx}(\tau)^2 \geq$ $-2C_{Q,x}(V_x)v_{TH,x}(\tau)V_x - C_{Q,x}(V_x)V_x^2$	$\varphi_x + \delta_x$
2(c)	S_{x2}	$i_{Lx}(\tau) < 0$	$\frac{1}{2}L_{TH,x}i_{Lx}(\tau)^2 \geq$ $2C_{Q,x}(V_x)v_{TH,x}(\tau)V_x + C_{Q,x}(V_x)V_x^2$	$\varphi_x + \pi - \delta_x$
2(d)	S_{x4}	$i_{Lx}(\tau) > 0$	$\frac{1}{2}L_{TH,x}i_{Lx}(\tau)^2 \geq$ $-2C_{Q,x}(V_x)v_{TH,x}(\tau)V_x + C_{Q,x}(V_x)V_x^2$	$\varphi_x - \delta_x$

is about to turn on. Consequently, a residual voltage ΔV_x remains across the drain-source of the MOSFET before it turns on.

In such a partial ZVS scenario, calculating ΔV_x is crucial for determining the $C_{OSS,x}$-related losses, which contribute to the total switching loss of the device. To obtain ΔV_x, the energy expressions must be revised. To illustrate the calculation process, we consider a specific switching transition: Case 1(a).

In case-1(a), S_{x1} and S_{x4} are turning-off while S_{x2} and S_{x3} are turning-on. Assuming incomplete-ZVS, the output capacitors $(C_{OSS,x})$ of $S_{x,1}$ and $S_{x,4}$ going to charge from 0 V to $(V_x - \Delta V_x)$ while the output capacitors of S_{x2} and S_{x3} will discharge from V_x to ΔV_x.

The total energy within the system at the start of the commutation is the sum of the energy stored in $L_{TH,x}$ due to inductor current $i_{Lx}(\tau)$ and the stored energy in C_{OSS} of S_{x1} and S_{x4}, which is represented in (2.69).

$$E_{inital} = \frac{1}{2}L_{TH,x}i_{Lx}(\tau)^2 + C_{E,x}(V_x)V_x^2. \tag{2.69}$$

Here, $C_{E,x}(V_x)$ is the energy-equivalent body capacitance due to V_x voltage appearing across it. $C_{E,x}(V_x)$ can be calculated as,

$$C_{E,x}(V_x) = \frac{2\int_0^{V_x}(\int_0^{v_{ds}} C_{oss}(y)dy) \cdot dv_{ds}}{V_x^2} \tag{2.70}$$

Further, at the end of commutation for a case with critical ZVS inductor current or at a valley switching, total energy stored in the body capacitors and the inductor is,

$$E_{final} = C_{E,x}(\Delta V_x)\Delta V_x^2 + C_{E,x}(V_x - \Delta V_x).(V_x - \Delta V_x)^2 \tag{2.71}$$

Also, observing Fig. 2.21, in this circuit configuration, the total energy sunk by the sources is,

$$E_{sunk} = \int_0^{\tau_c}\left[v_{TH,x}(\tau)i_{Lx}(\tau) - V_xI_x(\tau)\right]dt$$

$$= \int_0^{\tau_c}\left[v_{TH,x}(\tau)\left(2C_{OSS,x}(v_{ds,S_{x1}})\frac{dv_{ds,S_{x1}}}{dt}\right) - 0\right]dt$$

$$= \int_0^{V_x - \Delta V_x} 2v_{TH,x}(\tau)C_{OSS,x}(v_{ds,S_{x1}})dv_{ds,S_{x1}}$$

$$= 2C_{Q,x}(V_x - \Delta V_x)v_{th,x}(\tau)(V_x - \Delta V_x) \tag{2.72}$$

Now, from the law of energy conservation,

$$E_{final} = E_{sunk} + E_{initial} \tag{2.73}$$

Solving the equality given in (2.73), the remaining drain-source voltage of S_{x2} and S_{x3}, ΔV_x can be obtained. Similarly, for other switching transition cases, drain-source voltage at turn-on instant ΔV_x under incomplete ZVS action can be derived. Under hard-switching or no ZVS case, $\Delta V_x = V_x$ and for complete ZVS, $\Delta V_x = 0$.

Now, let us find out the total energy dissipated due to an incomplete ZVS transition in any switching leg. Suppose S_{x1} turns on while $C_{oss,S_{x1}}$ is still charged to ΔV_x voltage, which dissipates a certain amount of energy that can be derived by solving the energy balance of

$$E'_{final} + E'_{diss} = E'_{sunk} + E'_{initial}. \tag{2.74}$$

Before S_{x1} turns on, the energy within the switching leg system (C_{oss} of S_{x1} and S_{x3}) is

$$E'_{initial} = \frac{1}{2}C_{E,x}(\Delta V_x)\Delta V_x^2 + \frac{1}{2}C_{E,x}(V_x - \Delta V_x).(V_x - \Delta V_x)^2. \tag{2.75}$$

After S_{x1} turns on, C_{oss} of S_{x3} is charged to V_x, therefore, the final energy in the system is

$$E'_{final} = \frac{1}{2}C_{E,x}(V_x)V_x^2. \tag{2.76}$$

In order to charge the C_{oss} of S_{x3}, the remaining charge is

$$\Delta Q_{S_{x3}} = C_{Q,x}(V_x).V_x - C_{Q,x}(V_x - \Delta V_x).(V_x - \Delta V_x). \tag{2.77}$$

This amount of charge has to be taken from the dc source V_x. Thus,

$$E'_{sunk} = \Delta Q_{S_{x3}} V_x. \tag{2.78}$$

Hence, the total dissipated energy due to the partial ZVS transition of S_{x1} can be derived using (2.74) as,

$$\begin{aligned}E'_{diss} &= \frac{1}{2}C_{E,x}(\Delta V_x)\Delta V_x^2 + \frac{1}{2}C_{E,x}(V_x - \Delta V_x).(V_x - \Delta V_x)^2 + \Delta Q_{S_{x3}} V_x - \frac{1}{2}C_{E,x}(V_x)V_x^2 \\ &= \left[\frac{1}{2}C_{E,x}(\Delta V_x)\Delta V_x^2\right] + \left[\Delta Q_{S_{x3}} V_x\right] - \left[\frac{1}{2}C_{E,x}(V_x)V_x^2 - \frac{1}{2}C_{E,x}(V_x - \Delta V_x).(V_x - \Delta V_x)^2\right]\end{aligned} \tag{2.79}$$

In the expression of (2.78), the first term denotes the energy dissipated at S_{x1} when S_{x1} turns on. The second term represents the energy provided by source during turn-on of S_{x1} and the third term showcases the share of the energy provided by source that is stored in S_{x3}.

For complete ZVS with $\Delta V_x = 0$, (2.79) yields $E'_{diss} = 0$. For hard-switching with $\Delta V_x = V_x$, the dissipated energy will become, $E'_{diss} = C_{Q,x}(V_x).V_x^2$.

2.5.3 Solutions for Charge-Based ZVS Conditions

Another approach of evaluating the ZVS criteria for the individual switching legs of a MAB converter is detailed here, where the equivalent capacitance of the switching leg is kept under consideration rather individual capacitances of the high and low side MOSFETs [7]. As the high side and low side MOSFETs placed on the same switching leg exhibit complementary switching, achieving ZVS in one of them confirms the same for the other. Consequently, the analysis here focuses on the commutation process and ZVS characteristics of S_{x1} and S_{x2}, shown in Fig. 2.19. Theoretically, three following switching conditions are considered: S_{x1} turns on alone, S_{x2} turns on alone, (case-1: when $\delta_x >$ 0) and S_{x1} and S_{x4} turn ON simultaneously (case-2: when $\delta_x = 0$). The corresponding commutation process during the dead-band is illustrated in Fig. 2.22.

Analyzing ZVS operation under case-1, we focus on the transient process during the turn-on of S_{x1}. Figure 2.21 illustrates the current path during dead time. Following the turn-off of S_{x3}, inductor $L_{th,x}$ resonates with junction capacitances $C_{OSS,S_{x1}}$ and $C_{OSS,S_{x3}}$. Inductor current (i_{Lx}) charges $C_{OSS,S_{x3}}$ and discharges $C_{OSS,S_{x1}}$. The voltage across $C_{OSS,S_{x3}}$ begins to rise, while the voltage across $C_{OSS,S_{x1}}$ decreases that leads v_x to rise from 0V to dc link voltage V_x, as depicted in Fig. 2.22. As $C_{OSS,S_{x1}}$ voltage reaches 0, a small negative i_{Lx} flows through the body diode of S_{x1}, before its channel turns on, signifying the achievement of ZVS. The critical ZVS current ($I_{Lx,1,crit}$) is defined as the current that results in i_{Lx} being exactly 0 at the end of dead time (T_{dead}), representing the minimum i_{Lx} value necessary for achieving ZVS in the corresponding switch.

Due to the strong nonlinearity of the junction capacitance, accurately describing the v_x waveform during dead time is challenging. A linear approximation proposed in [8] and [9] that piecewise linearizes v_x, assuming it changes instantly from 0 to V_x when actual v_x equals $V_x/2$. T_A represents the time when $v_x = V_x/2$, and the equivalent i_{Lx} is linearized

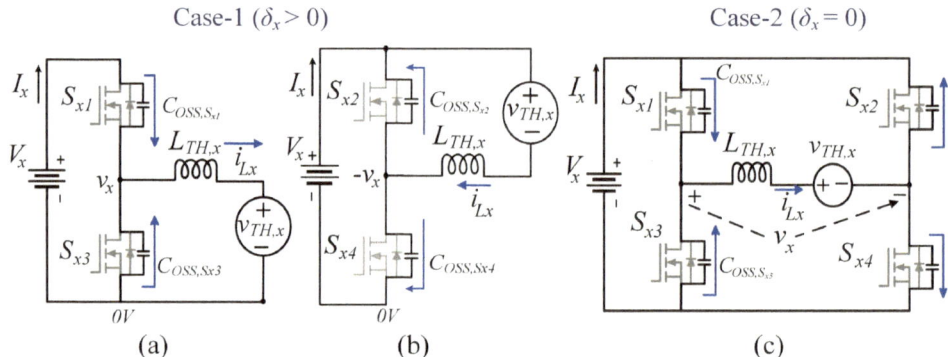

Fig. 2.22 Different switching commutation scenarios in a TAB H-bridge: **a** S_{x1} turns on alone; **b** S_{x2} turns on alone; and **c** S_{x1} and S_{x4} turn ON simultaneously

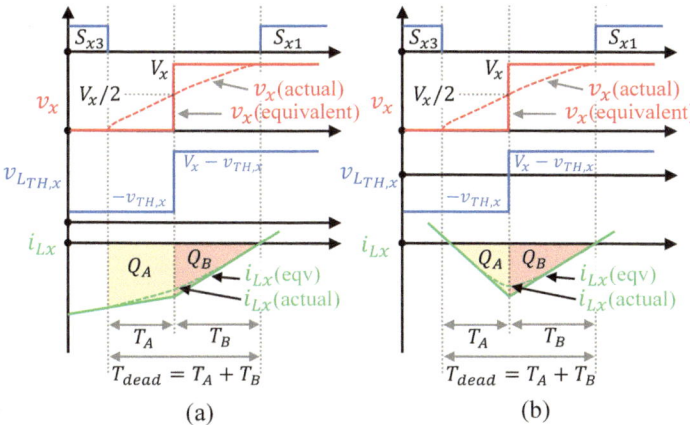

Fig. 2.23 Actual and approximated ZVS transient waveforms of the TAB port voltage and current when **a** $v_{TH,x} < 0$ and **b** $V_x > v_{TH,x}(t) > 0$

into two pieces with T_A as the middle point, as depicted in Fig. 2.23. This linearized i_{Lx} during the dead time is mathematically expressed as (2.80).

$$i_{Lx}(t) = \begin{cases} i_{Lx}(T_A) - \frac{v_{TH,x}}{L_{TH,x}}(t - T_A), t \in [0, T_A) \\ i_{Lx}(T_A) + \frac{(V_x - v_{TH,x})}{L_{TH,x}}(t - T_A), t \in [T_A, T_{dead}) \end{cases}. \tag{2.80}$$

Further, considering $i_{Lx}(T_{dead}) = 0$ in Fig. 2.22a, $T_B (= T_{dead} - T_A)$ can be derived as, $\frac{-i_{Lx}(T_A)L_x}{V_x - V_{TH,x}}$.

To achieve ZVS, it is imperative to completely discharge the stored charges $C_{OSS,S_{x1}}$ and charge $C_{OSS,S_{x3}}$ through the inductor current, i_{Lx}. The shaded region in Fig. 2.23 represents the total charges liberated by i_{Lx}. This region is partitioned into Q_A and Q_B, signifying the conveyed charges during the v_x transition from 0 to $V_x/2$ and $V_x/2$ to 0, respectively, which are calculated using (2.81).

$$\begin{cases} Q_A = \int_0^{T_A} i_{Lx}(t).dt = |i_{Lx}(T_A)|T_A - \frac{v_{TH,x}T_A^2}{2L_{TH,x}} \\ Q_B = \int_{T_A}^{T_{dead}} i_{Lx}(t).dt = \frac{|i_{Lx}(T_A)|^2 L_{TH,x}}{2(V_x - v_{TH,x})}. \end{cases} \tag{2.81}$$

Simultaneously, the C_{oss} of the MOSFETs experiences nonlinear changes concerning change in drain-source voltage v_{DS}. The stored charges in the C_{oss} necessitate calculation and can be derived by integrating the nonlinear C_{oss}-v_{DS} plot available in the MOSFET's datasheet, as shown in Fig. 2.24. The shaded area in Fig. 2.24 represents the total charge stored in the $C_{OSS,S_{x1}}$ and $C_{OSS,S_{x3}}$ pair that can be divided into two parts: $Q_{a(V_x)}$ and $Q_{b(V_x)}$, depending on v_x transition from 0 to $V_x/2$ and $V_x/2$ to V_x. Due to $C_{OSS,S_{x1}}(v_x) =$

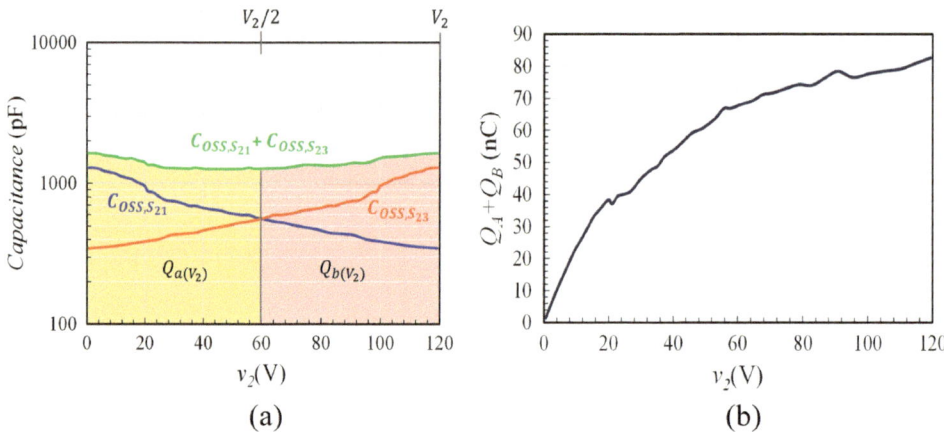

Fig. 2.24 Variation of **a** C_{oss} and **b** its stored charge with varying v_{DS} for EPC 2215 GaN MOSFET

$C_{OSS,S_{x3}}(V_x - v_x)$, the relation between Q_a and Q_b is formed as

$$Q_{a(V_x)} = \int_0^{\frac{V_x}{2}} C_{OSS,S_{x1}}(v_x).dv_x + \int_{V_x}^{\frac{V_x}{2}} C_{OSS,S_{x3}}(V_x - v_x).dv_x$$

$$= \int_{\frac{V_x}{2}}^{V_x} C_{OSS,S_{x1}}(v_x).dv_x + \int_{\frac{V_x}{2}}^{0} C_{OSS,S_{x3}}(V_x - v_x).dv_x = Q_{b(V_x)}. \qquad (2.82)$$

The prerequisite for achieving ZVS is that the conveyed charges in i_{Lx} must surpass the capacitance stored charges. This results in the constraint

$$Q_A = Q_B \geq Q_{a(V_x)} = Q_{b(V_x)}. \qquad (2.83)$$

Consequently, based on Eqs. (1.70)–(1.72), the critical ZVS current $I_{Lx,1,crit}$ of S_{x1} (under condition of Fig. 2.23a) is derived as,

$$I_{Lx,1,crit} = -\sqrt{2Q_{B(V_x)}\big[V_x - v_{TH,x}(t)\big]/L_{TH,x}} \qquad (2.84)$$

For any switching current i_{Lx} lower than $I_{Lx,1,crit}$, S_{x1} will undergo soft-switching.

In a similar way as stated above, the critical ZVS currents for other switching transitions (as depicted in Fig. 2.22) are formulated and highlighted in Table 2.9. Therefore, depending on the operating state of the TAB defined by the status of the control variables $(\varphi_2, \varphi_3, \delta_1, \delta_2, \delta_3)$ and its circuit parameters such as dc link port voltages, port inductances and switching device parameters, the required ZVS currents can be identified online.

Table 2.9 Summarized ZVS current requirements for Port-x in a MAB

Case	δ_x	Leg undergoing turn-on event	Min required Switch Currents for ZVS	Switching time ($\omega_s t$)
1	+ve	Leg-1	$I_{Lx,1,crit} = \begin{cases} -\sqrt{2Q_{B(V_x)}[V_x - v_{TH,x}(t)]/L_{TH,x}}; & if V_x > 0 > v_{TH,x}(t) \\ -max\left[\sqrt{2Q_{B(V_x)}[V_x - v_{TH,x}(t)]/L_{TH,x}}, \sqrt{2Q_{A(V_x)}v_{TH,x}(t)/L_{TH,x}}\right]; & if V_x > v_{TH,x}(t) \\ 0; & if v_{TH,x}(t) > V_x > 0 \end{cases}$	$\varphi_x + \delta_x$
		Leg-2	$I_{Lx,2,crit} = \begin{cases} 0; & if V_x > 0 > v_{TH,x}(t) \\ max\left[\sqrt{2Q_{A(V_x)}[V_x - v_{TH,x}(t)]/L_{TH,x}}, \sqrt{2Q_{B(V_x)}v_{TH,x}(t)/L_{TH,x}}\right]; & if V_x > v_{TH,x}(t) \\ \sqrt{2Q_{B(V_x)}v_{TH,x}(t)/L_{TH,x}}; & if v_{TH,x}(t) > V_x > 0 \end{cases}$	$\pi + \varphi_x - \delta_x$
2	0	Any	$I_{Lx,1,crit} = -I_{Lx,2,crit} = \begin{cases} -\sqrt{4Q_{B(V_x)}[V_x - v_{TH,x}(t)]/L_{TH,x}}; & if V_x > 0 > v_{TH,x}(t) \\ -\sqrt{4Q_{B(V_x)}[V_x + v_{TH,x}(t)]/L_{TH,x}}; & if V_x > v_{TH,x}(t) > 0 \\ 0; & if v_{TH,x}(t) > V_x > 0 \end{cases}$	φ_i

2.6 Generalized Modeling of n-Port MAB Considering Circuit Non-idealities

In the previous sections of this chapter, converters within the MAB family—such as DAB, TAB, and generalized n-port MAB—have been modeled in either the time domain or frequency domain, utilizing ideal active or passive elements within the circuit. However, in practical scenarios, the switching semiconductor devices such as MOSFETs within the H-bridges exhibit non-zero on-state resistances. Other non-idealities in the circuit include the AC winding resistances of line inductors, transformers, and the equivalent series resistance (ESR) of DC blocking capacitors and DC link capacitors. These non-ideal resistive elements cause resistive voltage drops and conduction losses in the circuit, leading to inaccuracies in power transfer calculations when using expressions derived from ideal models. Moreover, the assumption that a DC blocking capacitor acts as a simple short circuit can fail with changing operational frequencies, due to its variable impedance. To address these challenges, this section presents an improved unified modeling approach that captures the effects of these non-idealities in an n-port MAB converter. Furthermore, using such modeling methodology, most of the transformer isolated H-bridge based non-resonant or resonant converters comprising of any number of ports can be precisely modeling while accounting for the resistive non-idealities in their circuits.

Figure 2.25 highlights a non-resonant or resonant MAB converter with **n** active H-bridges magnetically coupled together through a n-winding high-frequency (HF) transformer having a turns ratio of $n_1, n_2, \ldots n_n$. Port-i ($i \in [1, n]$) of such a MAB design has an associated dc link voltage $V_{i'}$, line impedance $Z_{i'}$, bridge output voltage $v_{i'}$, and winding current $i_{i'}$. Each of the MAB port has its associated line/resonant inductor $L_{i'}$ and a dc blocking or resonant capacitor $C_{i'}$ depending on the circuit topology. As a part of the non-idealities, each of the MOSFETs connected at port-i H-bridge has an on-state resistance of $R_{on,i'}$; the ac winding ac resistance of the inductor and transformer corresponding to ith bridge is considered to be $R_{ac,winding,i'}$; the ESR of the dc blocking or resonant capacitor connected in series with the line inductor is $R_{ESR,C_i'}$. Thus, the line impedance can be generalized as,

$$Z_i' = R_i' + j.2\pi k f_{sw} L_i' - j.\frac{1}{2\pi k f_{sw} C_i'} \tag{2.85}$$

where, $L_{i'}$, $C_{i'}$, and $R_{i'}$ represent the line inductor, capacitor and lumped resistance of the path for port-i, respectively. Here the lumped resistance $R_{i'}$ can be calculated as,

$$R_i' = 2R_{on,i}' + R_{ac,winding,i}' + R_{ESR,C_i}'. \tag{2.86}$$

As discussed previously, the degrees of freedom to control power flow in an MAB topology are defined by the intra-bridge leg phase shift or $v_{i'}$ duty ratios (δ_i), the inter-bridge phase differences (φ_i, and $\varphi_1 = 0$) or the phase differences between the different

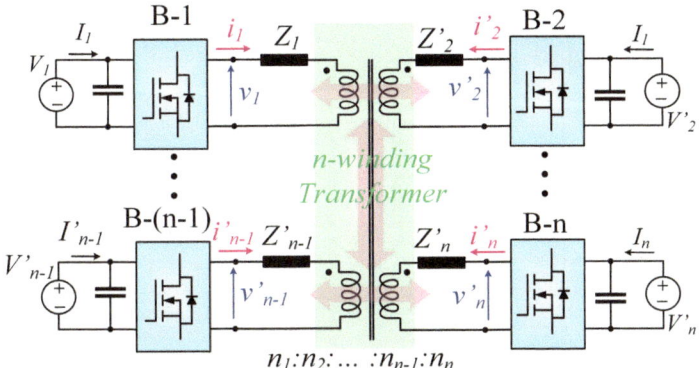

Fig. 2.25 Circuit schematic of n-port MAB

bridge voltages and the switching frequency (f_{sw}). The range of the phase-duty variables are defined as: $\delta_i \in [0, \pi/2]$; $\varphi_i \in [-\pi/2, \pi/2]$. Thus, for a n-port MAB, total number of possible control variables is $2n$, including (n-1) phase-shifts, n duty parameters and f_{sw}. Now, depending on the actual circuit topology—resonant or non-resonant, the modulation strategy can be defined as phase-shift modulation (PSM), hybrid phase-duty modulation, pulse frequency modulation, or their combinations.

Furthermore, the port-1 referred Y model of the MAB circuit is illustrated in Fig. 2.25, where the modified circuit parameters are: $Z_i = Z_i' \left(\frac{n_i}{n_1}\right)^2$; $v_i = v_i' \left(\frac{n_1}{n_i}\right)$; $i_i = i_i' \left(\frac{n_i}{n_1}\right)$ and $Z_m = j.2\pi k f_{sw} L_m$, where L_m is the magnetizing inductance of the transformer referred to the primary side or port-1. Depending on the actual circuit topology of the converter under the MAB family, the definitions of the respective port impedance Z_i can vary, which is highlighted in Table 2.10. Thus, it can be concluded that using our generalized MAB model any converter mentioned in the list of Table 2.10 can be analyzed without requiring separate modeling methods individually.

For ease of analysis and to utilize the superposition theorem, we first consider the power flow between any two ports (i and j) of the MAB circuit. The Thevenin equivalent circuit between port-i and port-j is derived and depicted in Fig. 2.26b.

The Thevenin equivalent impedance $Z_{th,ij}$ can be equated as:

$$Z_{th,ij} = \left(\frac{1}{Z_{m,k}} + \sum_{a \neq i,j}^{n} \frac{1}{Z_{a,k}} \right)^{-1} \qquad (2.87)$$

Using the laws of star-delta transformation, the Δ-circuit model impedances are derived and given in (2.88).

Table 2.10 Possible dc-dc converter topologies modeled using generalized MAB modeling methodology

MAB non-resonant or resonant topology		Impedances	Description of passive elements (referred to primary)	Impedance tank
Two port	DAB	$Z_1 = R_1 + j.2\pi kf_{sw}L_1 - j.\frac{1}{2\pi kf_{sw}C_1}$ $Z_2 = R_2 + j.2\pi kf_{sw}L_2 - j.\frac{1}{2\pi kf_{sw}C_2}$ $Z_m = j.2\pi kf_{sw}L_m$	L_1, L_2 = primary and secondary side line inductors C_1, C_2 = primary and secondary side dc blocking capacitors if any L_m = magnetizing inductance of the transformer	
	SR-DAB	$Z_1 = R_1 + j.2\pi kf_{sw}L_1 - j.\frac{1}{2\pi kf_{sw}C_1}$ $Z_2 = R_2 + j.2\pi kf_{sw}L_2 - j.\frac{1}{2\pi kf_{sw}C_2}$ $Z_m = j.2\pi kf_{sw}L_m$	L_1, L_2 = primary and secondary side resonant inductors C_1, C_2 = primary and secondary side resonant capacitors if any L_m = magnetizing inductance of the transformer	
	CLLC	$Z_1 = R_1 + j.2\pi kf_{sw}L_1 - j.\frac{1}{2\pi kf_{sw}C_1}$ $Z_2 = R_2 + j.2\pi kf_{sw}L_2 - j.\frac{1}{2\pi kf_{sw}C_2}$ $Z_m = j.2\pi kf_{sw}L_m$	L_1, L_2 = primary and secondary side resonant inductors C_1, C_2 = primary and secondary side resonant capacitors L_m = magnetizing inductance of the transformer	
	LLC	$Z_1 = R_1 + j.2\pi kf_{sw}L_1 - j.\frac{1}{2\pi kf_{sw}C_1}$ $Z_2 = R_2$ $Z_m = j.2\pi kf_{sw}L_m$	L_1 = primary side resonant inductor C_1 = primary side resonant capacitor L_m = magnetizing inductance of the transformer	

(continued)

Table 2.10 (continued)

MAB non-resonant or resonant topology		Impedance tank	Description of passive elements (referred to primary)	Impedances
	CLL		C_1 = primary side resonant capacitor L_2 = secondary side resonant inductor L_m = magnetizing inductance of the transformer	$Z_1 = R_1 - j \cdot \dfrac{1}{2\pi k f_{sw} C_1}$ $Z_2 = R_2 + j \cdot 2\pi k f_{sw} L_2$ $Z_m = j \cdot 2\pi k f_{sw} L_m$
Three port	TAB		L_1, L_2, L_3 = primary, secondary, and tertiary side line inductors C_1, C_2, C_3 = primary, secondary, and tertiary side dc blocking capacitors if any L_m = magnetizing inductance of the transformer	$Z_1 = R_1 + j.2\pi k f_{sw} L_1 - j \cdot \dfrac{1}{2\pi k f_{sw} C_1}$ $Z_2 = R_2 + j.2\pi k f_{sw} L_2 - j \cdot \dfrac{1}{2\pi k f_{sw} C_2}$ $Z_3 = R_3 + j.2\pi k f_{sw} L_3 - j \cdot \dfrac{1}{2\pi k f_{sw} C_3}$ $Z_m = j.2\pi k f_{sw} L_m$
	C3L3		L_1, L_2, L_3 = primary, secondary, and tertiary side line inductors C_1, C_2, C_3 = primary, secondary, and tertiary side dc blocking capacitors if any L_m = magnetizing inductance of the transformer	$Z_1 = R_1 + j.2\pi k f_{sw} L_1 - j \cdot \dfrac{1}{2\pi k f_{sw} C_1}$ $Z_2 = R_2 + j.2\pi k f_{sw} L_2 - j \cdot \dfrac{1}{2\pi k f_{sw} C_2}$ $Z_3 = R_3 + j.2\pi k f_{sw} L_3 - j \cdot \dfrac{1}{2\pi k f_{sw} C_3}$ $Z_m = j.2\pi k f_{sw} L_m$
Four port	QAB		L_1, L_2, L_3, L_4 = line inductors connected at port-1 to 4 C_1, C_2, C_3, C_4 = dc blocking capacitors connected at port-1 to 4 L_m = magnetizing inductance of the transformer	$Z_1 = R_1 + j.2\pi k f_{sw} L_1 - j \cdot \dfrac{1}{2\pi k f_{sw} C_1}$ $Z_2 = R_2 + j.2\pi k f_{sw} L_2 - j \cdot \dfrac{1}{2\pi k f_{sw} C_2}$ $Z_3 = R_3 + j.2\pi k f_{sw} L_3 - j \cdot \dfrac{1}{2\pi k f_{sw} C_3}$ $Z_4 = R_4 + j.2\pi k f_{sw} L_4 - j \cdot \dfrac{1}{2\pi k f_{sw} C_4}$ $Z_m = j.2\pi k f_{sw} L_m$

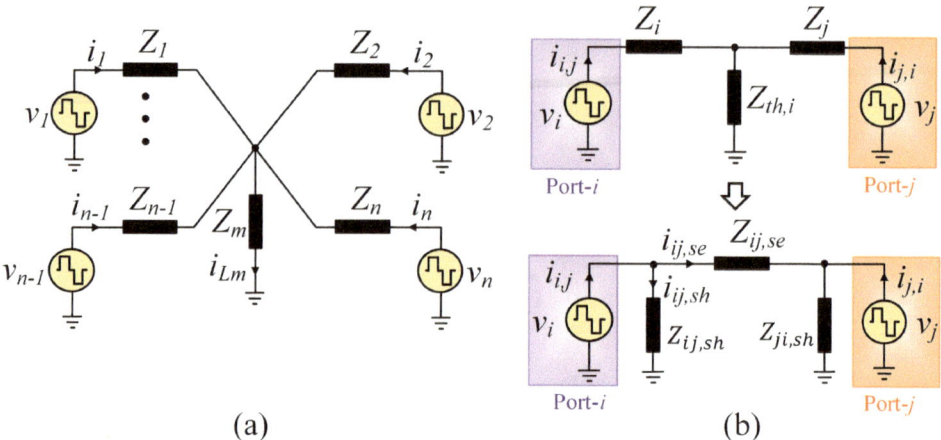

Fig. 2.26 MAB Converter Models: **a** Y-model; **b** equivalent circuit derivation considering power flow between port-i and j

$$Z_{ij,se} = Z_i + Z_j + Z_i Z_j \left(\frac{1}{Z_m} + \sum_{\substack{a=1 \\ a \neq i,j}}^{n} \frac{1}{Z_a} \right); i \neq j$$

$$Z_{ij,sh} = Z_i + \left(\frac{1}{Z_m} + \sum_{\substack{a=1 \\ a \neq i,j}}^{n} \frac{1}{Z_a} \right)^{-1} \left(1 + \frac{Z_i}{Z_j} \right); i \neq j$$ (2.88)

$$Z_{ji,sh} = Z_j + \left(\frac{1}{Z_m} + \sum_{\substack{a=1 \\ a \neq i,j}}^{n} \frac{1}{Z_a} \right)^{-1} \left(1 + \frac{Z_j}{Z_i} \right); i \neq j$$

Once the equivalent circuit is modeled, the critical step is to synthesize the winding current (i_i) shapes, its RMS, and switching instant values. This is achieved through a GHA based frequency domain analysis, as presented earlier in this chapter. The quasi-square shaped port voltage (v_i) is expanded in a Fourier series with odd order harmonics up to nth order, as given in (2.89), where $d_{ki} = \cos(k\delta_i)$; 'k' is the order of the harmonic, and $\omega = 2\pi f_{sw}$.

$$v_i(t) = \frac{4V_i}{\pi} \cdot \sum_{k=1,3,\ldots}^{\infty} \frac{1}{k} d_{ki} \sin\{k(\omega t - \varphi_i)\}$$ (2.89)

Following this, the current flowing from port i to port j through their corresponding series impedance $Z_{ij,se}$ can be deduced from ac circuit analysis knowledge and is shown in (2.90).

$$i_{ij,se}(t) = \sum_{k=1,3,\ldots}^{\infty} \left[\frac{|v_{i,k}| \cdot \sin\{k(\omega t - \varphi_i) - \angle Z_{ij,se,k}\} - |v_{j,k}| \cdot \sin\{k(\omega t - \varphi_j) - \angle Z_{ij,se,k}\}}{|Z_{ij,se,k}|} \right],$$ (2.90)

where $|v_{i,k}| = \frac{4V_i d_{k,i}}{\pi k}$.

Similarly, the current flowing through the shunt impedance $Z_{ij,sh}$ for port-i is derived as:

$$i_{ij,sh}(t) = \sum_{k=1,3,...}^{\infty} \left[\frac{|v_{i,k}| \cdot \sin\{k(\omega t - \varphi_i) - \angle Z_{ij,sh,k}\}}{|Z_{ij,sh,k}|} \right] \tag{2.91}$$

Using KCL, the current sourced by port-i within the equivalent circuit formed with port i and j is derived as:

$$i_{i,j}(t) = i_{ij,sh}(t) + i_{ij,se}(t). \tag{2.92}$$

For an MAB converter with n-ports, $(n-1)$ such equivalent circuits can be formed. Therefore, the total current sourced by any port-i, i.e., the winding current corresponding to port-i, is formulated using the superposition theorem:

$$i_i(t) = \sum_{j \neq i}^{n} i_{ij,sh}(t) + i_{ij,se}(t)$$

$$= \sum_{k=1,3,...}^{\infty} \left[A_{i,k} \cos(k\omega t) + B_{i,k} \sin(k\omega t) \right] \tag{2.93}$$

Here, the harmonic order dependent quantities $A_{i,k}$, and $B_{i,k}$ are represented in (2.94) and (2.95).

$$A_{i,k} = \sum_{j \neq i}^{n} \left[\frac{|v_{j,k}|}{|Z_{ij,se,k}|} \cdot \sin(k\varphi_j + \angle Z_{ij,se,k}) - \frac{|v_{i,k}|}{|Z_{ij,se,k}|} \cdot \sin(k\varphi_i + \angle Z_{ij,se,k}) \right.$$

$$\left. - \frac{|v_{i,k}|}{|Z_{ij,sh,k}|} \cdot \sin(k\varphi_j + \angle Z_{ij,sh,k}) \right] \tag{2.94}$$

$$B_{i,k} = \sum_{j \neq i}^{n} \frac{|v_{i,k}|}{|Z_{ij,se,k}|} \cdot \cos(k\varphi_i + \angle Z_{ij,se,k}) - \frac{|v_{j,k}|}{|Z_{ij,se,k}|} \cdot \cos(k\varphi_j + \angle Z_{ij,se,k})$$

$$+ \frac{|v_{i,k}|}{|Z_{ij,sh,k}|} \cdot \cos(k\varphi_j + \angle Z_{ij,sh,k}) \right] \tag{2.95}$$

Moreover, by employing the continuous time current expression of i_i, the RMS value of the current can be determined as:

$$i_{i,RMS}^2 = \frac{1}{2} \sum_{k=1,3,..}^{\infty} \frac{1}{k^4} [A_{i,k}^2 + B_{i,k}^2] \tag{2.96}$$

This expression can be directly applied to compute the winding current RMS values in any MAB converter, which is crucial for quantifying the conduction losses in the MOSFETs and transformer windings.

The total power sourced by port-i of the MAB converter is synthesized from first principles:

$$P_i = \sum_{k=1,3,\ldots}^{\infty} v_{i,k,RMS} \cdot i_{i,k,RMS} \cdot \cos\left\{ k\varphi_i + \tan^{-1}\left(\frac{B_{i,k}}{A_{i,k}}\right) \right\} \qquad (2.97)$$

where $v_{i,k,RMS} = \frac{2\sqrt{2}V_i d_{k,i}}{\pi k}$ and $i_{i,k,RMS} = \sqrt{\frac{A_{i,k}^2 + B_{i,k}^2}{2}}$.

This comprehensive approach allows for precise modeling of power flow and winding currents in a practical, loss-inclusive MAB converter.

2.7 Conclusions

This chapter presents an extensive circuit analysis along with time-domain and frequency-domain modeling techniques for 2-port, 3-port, and generalized n-port converters within the MAB family. It is demonstrated that theoretical modeling of the MAB network using a frequency-domain Generalized Harmonic Approximation (GHA) technique can effectively bypass the time-intensive, mode-dependent calculations required in time-domain analyses. Moreover, unified expressions for the instantaneous high-frequency winding currents, their RMS values, bridge voltages, and inter-port power flow are formulated as functions of the operating port voltages and control variables such as inter- and intra-bridge phase shifts and switching frequency. The ZVS criteria for the H-bridge switches in a MAB are also derived using a generic port-equivalent model. These developments serve as a foundation for the subsequent chapters, where converter losses will be modeled as functions of the operating control variables, and the control parameters will be optimized to ensure efficient converter performance.

References

1. M. H. Kheraluwala, R. W. Gascoigne, D. M. Divan, and E. D. Baumann, "Performance characterization of a high-power dual active bridge dc-todc converter," IEEE Trans. Ind. Appl., vol. 28, no. 6, pp. 1294–1301, Nov./Dec. 1992.
2. Chuanhong Zhao and J. W. Kolar, "A novel three-phase three-port UPS employing a single high-frequency isolation transformer," *2004 IEEE 35th Annual Power Electronics Specialists Conference (IEEE Cat. No.04CH37551)*, Aachen, Germany, 2004, pp. 4135–4141 Vol.6, https://doi.org/10.1109/PESC.2004.1354730.

3. A. K. Jain and R. Ayyanar, "Pwm control of dual active bridge: Comprehensive analysis and experimental verification," in *IEEE Transactions on Power Electronics*, vol. 26, no. 4, pp. 1215–1227, April 2011, https://doi.org/10.1109/TPEL.2010.2070519.
4. S. Dey and A. Mallik, "Switching Network Loss Minimization Through Multivariable Modulation in a Multiactive Bridge Converter," in *IEEE Transactions on Industrial Electronics*, vol. 70, no. 11, pp. 10833–10847, Nov. 2023, https://doi.org/10.1109/TIE.2022.3225806.
5. S. Dey and A. Mallik, "An Online-Optimized ZVS-Current Tracked Soft-Switching Modulation for Triple Active Bridge Converter," in *IEEE Transactions on Power Electronics*, https://doi.org/10.1109/TPEL.2024.3429278.
6. S. Zou, J. Lu, and A. Khaligh, "Modeling and control of a triple active bridge converter," IET Power Electron., vol. 13, no. 5, pp. 961–969, Apr. 2020.
7. Z. Wang, C. Li, J. Liu, and Z. Zheng, "Influence of junctioncapacitance and dead-time on dual-active-bridge actual soft-switchingrange: Analytic analysis and solution," IEEE Trans. Power Electron., vol. 38, no. 5, pp. 6157–6168, May 2023, https://doi.org/10.1109/TPEL.2022.3228338.
8. B. Liu, P. Davari, and F. Blaabjerg, "Nonlinear coss−VDS profile based ZVS range calculation for dual active bridge converters," IEEE Trans. Power Electron., vol. 36, no. 1, pp. 45–50, Jan. 2021.
9. J. Everts, F. Krismer, J. Van den Keybus, J. Driesen, and J. W. Kolar, "Charge-based ZVS soft switching analysis of a single-stage dual active bridge AC-DC converter," in Proc. IEEE Energy Convers. Congr. Expo., 2013, pp. 4820–4829.

Switching Modulator Optimization in DAB DC-DC and DC-AC Converters

3

3.1 Introduction of DAB Converters and Their Modeling Constraints

The DAB topology (see Fig. 3.1) consists of two active full bridges that apply quasi-square or square wave voltages to the transformer primary and secondary windings connected between them. Introducing phase shifts between these voltage waveforms controls power transfer between the H-bridges and is known as single phase shift strategy (SPS) based modulation [1]. Studies show that adding duty modulation to input or/and output side bridge voltages as additional control parameters offers advantages such as reducing transformer winding current peaks and RMS and extending the zero-voltage switching (ZVS) boundary [2], known as dual/triple phase shift strategy (DPS/TPS), respectively. Various research efforts have focused on improving DC-DC DAB converter operation using SPS, DPS, and TPS strategies [3].

Despite these advancements, current state-of-the-art (SOA) DAB designs face several challenges:

(a) Existing time-domain steady-state models of DC-DC DAB converters analyze circuit operation under various modes arising from different voltage levels at transformer input and output sides, and differing modulation parameter values such as phase shifts between active half-bridge cells. There is a need for a simpler, unified model that is independent of operating modes to accurately calculate instantaneous and RMS values of circuit voltages and currents with reduced computational effort and time.

(b) A few frequency-domain DAB models available use fundamental harmonic approximation techniques for voltage and current waveforms to minimize losses by reducing circulating currents or reactive components [4] of inductor currents. However, these

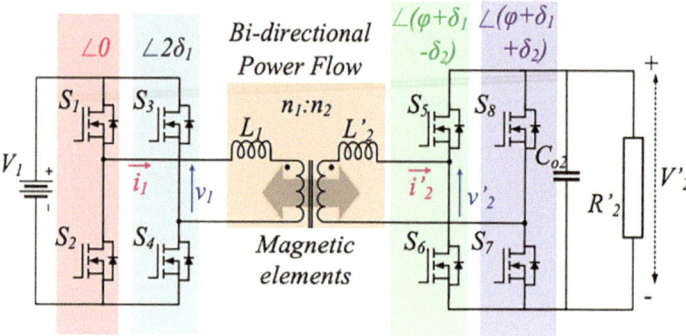

Fig. 3.1 DAB DC-DC converter topology and phase shifts of the individual half-bridge control signals

approximations may deviate significantly from actual current values across a wide converter operational range.

(c) Several studies have focused on improving DAB converter efficiency through multi-phase shift control strategies, formulating objective functions to minimize transformer winding RMS currents, peak currents, or total losses [5] using complex multi-operational zone time-domain analyses. However, there remains a lack of physical and analytical understanding regarding the relationship between the applicability and benefits of specific phase-shift control strategies and the converter's operating point specified by voltage gain and power level.

This chapter collectively addresses the technical gaps of state-of-the-art design optimization and holistically provides theoretical insight on multivariable modulator synthesis for DAB performance optimization.

3.2   Efficiency-Optimal Modulator Optimization in a DC-DC DAB Converter

The power transfer in a DAB converter fundamentally depends on the phase displacement between the H-bridge voltages (φ), their corresponding duty cycles (δ_1 and δ_2), terminal DC voltages and the power transfer line inductances. For controlling the power flow in a DAB DC-DC converter, only one control variable, primarily φ (SPS modulation), is enough and easy to implement as the solution of the power transfer equation i.e., (3.1) results in only one operating point in such case ($\delta_1 = \delta_2 = 0$).

$$P_{12} = \frac{4}{\pi^3 f_{sw}} \cdot \sum_{k=1}^{2m+1} \frac{1}{k^3} \left[\frac{V_1 V_2}{L} \cos(k\delta_2) \cos(k\delta_1) \sin(k\varphi) \right] \tag{3.1}$$

However, this operating point may not incur the highest possible efficiency of the converter for a particular power conversion mode for a wide gain and load range operation. The rest of the degrees of freedom, δ_1 and δ_2, are also needed to be efficiently utilized through a triple phase shift (TPS) modulation scheme in order to minimize the overall system losses while maintaining the same power flow condition. As an example, a multivariable design optimization problem is formulated where the TPS modulation strategy is employed in order to minimize the inductor RMS current that corresponds to the conduction loss in the system. Considering a DAB converter operation with a voltage gain of, $G = V_2/V_1$ at load Power 'P', the inductor RMS current is found to be a function of $V_1, G, L, \delta_1, \delta_2$ and φ and can be represented as an objective function per (3.2), which is to be minimized.

$$g(\delta_1, \delta_2, \varphi, G) = i_{L,rms}$$

$$= \sqrt{\frac{8V_1^2}{\pi^2\omega^2L^2} \cdot \sum_{k=1}^{2m+1} \frac{1}{k^4} \left[\begin{array}{c} \cos^2(k\delta_1) + G^2\cos^2(k\delta_2) \\ -2G\cos(k\delta_1)\cos(k\delta_2)\cos(k\varphi) \end{array} \right]} \qquad (3.2)$$

In this optimization problem, V_1, L, f_{sw} are kept fixed and treated as known constants whereas G and P are varied over a wide range. To supply a desired load power, the solutions of the control variable set $(\delta_1, \delta_2, \varphi)$ can be determined by solving the following power flow constraint (3.3).

$$PE(\delta_1, \delta_2, \varphi, P, G) = P_{12} - P = \frac{4}{\pi^3 f_{sw} LG} \cdot \sum_{k=1}^{2m+1} \frac{1}{k^3}[\cos(k\delta_2)\cos(k\delta_1)\sin(k\varphi)] - P = 0$$

$$(3.3)$$

The objective of the study is to find the optimum control variable sets for the TPS DAB operation that leads to minimum $i_{L,rms}$ while satisfying (3.3) for a wide variation in G and P_{out}. Similar optimization problems are formulated for the other modulation strategies as well such as double phase shift-1 (DPS-1), DPS-2 and SPS, which are essentially subsets of the TPS modulation only. In DPS-1, the secondary bridge is operated at 50% duty i.e., $\delta_2 = 0$. Likewise, the primary bridge is operated at 50% duty in DPS-2 modulation i.e., $\delta_1 = 0$.

To understand the quantitative effects of these modulation techniques on the converter power loss profile, a DAB example with the specifications mentioned in Table 3.1 is considered.

Now, all the formulated optimization problems corresponding to each modulation scheme are simultaneously solved for three different load power levels, i.e., 50, 200 and 400 W while the voltage gain is varied from 0.5 to 2. Figure 3.2a–c represent the optimized objective function, i.e., $i_{L,rms}$ value under the possible modulation strategies with a variation in the output voltage gain for 50 W, 200 W and 400 W load power, respectively.

Table 3.1 Design specifications of a DAB example

Circuit parameters	Values
Input DC voltage (V_1)	160 V
Output DC Voltage (V'_2)	90–150, 120 V (nominal)
Output Rated Power (P_{out})	400 W
Transformer Turns Ratio (N_1:N_2)	7:5
Magnetizing Inductance (L_m)	268 μH
Total Leakage Inductances (L_1, L_2)	16 and 15 μH
Output Capacitor (C_{o2})	86 μF
Switching frequency (f_{sw})	100 kHz

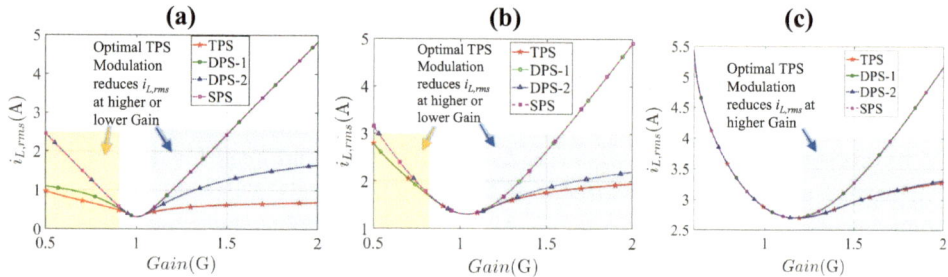

Fig. 3.2 Variation of $i_{L,rms}$ with changing voltage gain (G) under all the optimized modulation control strategies for **a** P_{out} = 50 W; **b** P_{out} = 200 W and **c** P_{out} = 400 W

It can be inferred from the plots that the utilization of conventional SPS modulation strategy results in optimum operating condition for a DAB converter when the converter operates at the vicinity of unity voltage gain. For a low or high gain converter operation, the employment of the TPS control is essential to maintain a lower $i_{L,rms}$ profile. Also, it is important to notice that as the load power decreases, the difference in the corresponding $i_{L,rms}$ currents under the optimized the TPS and SPS modulation schemes increases for a particular non-unity gain condition. Thus, the optimal TPS based control becomes more necessary when the converter is operating at light load mode with non-unity voltage gain condition. The reduction in $i_{L,rms}$ current under DPS-1, DPS-2 and TPS control schemes compared to the conventional SPS modulation are presented in Table 3.2 for a combination of three different load power (50, 200 and 400 W) and three separate output voltage modes (90, 120 and 150 V). The optimal modulation schemes for each power conversion mode to ensure minimum conduction loss while employing least number of control variables are marked in green color. Figures 3.3 and 3.4 represent the optimized $i_{L,rms}$ currents and the corresponding control variable sets (δ_1, φ_2, φ_3) under the optimized TPS and conventional SPS modulation techniques for a wide load power and gain variation in a 2D space. It can be observed that for a small operating area {G, P_{out}} at high load power and below unity voltage gain operation, the SPS modulation technique becomes the optimal control strategy, while the rest operating points require the TPS control in order to achieve the best possible converter efficiency.

The MATLAB scripts developed to generate the 2D and 3D plots of the DAB inductor current and engaged control variables are attached in the appendix section of this book.

Table 3.2 Inductor RMS current comparison between the DAB modulation strategies

V_{in}	V_{out}	Gain (G)	P_{out}	SPS modulation $(\delta_1, \delta_2, \varphi)$	$i_{L,rms}$	Optimal DPS-1 modulation $(\delta_1, \delta_2, \varphi)$	$i_{L,rms}$	% Reduction in $i_{L,rms}$ compared to SPS	Optimal DPS-2 modulation $(\delta_1, \delta_2, \varphi)$	$i_{L,rms}$	% Reduction in $i_{L,rms}$ compared to SPS	Optimal TPS modulation $(\delta_1, \delta_2, \varphi)$	$i_{L,rms}$	% Reduction in $i_{L,rms}$ compared to SPS
[V]	[V]		[W]		[A]		[A]	[%]		[A]	[%]		[A]	[%]
160	90	0.788	50	(0,0, 0.0762)	1.075	(0.53,0, 0.11)	0.79	26.51	(0,0, 0.0762)	1.075	0.00	(0.906, 0.73, 0.176)	0.625	41.86
			200	(0,0, 0.333)	1.824	(0.317,0, 0.374)	1.776	2.63	(0,0, 0.333)	1.824	0.00	(0.317,0, 0.374)	1.776	2.63
			400	(0,0, 0.0797)	3.582	(0,0, 0.797)	3.582	0.00	(0,0, 0.797)	3.582	0.00	(0, 0.73, 0.797)	3.582	0.00
160	120	1.050	50	(0,0, 0.057)	0.3918	(0,0, 0.057)	0.3918	0.00	(0,0.11, 0.0603)	0.3866	1.33	(0.177, 0.24, 0.066)	0.3833	2.17
			200	(0,0, 0.243)	1.31	(0,0, 0.243)	1.31	0.00	(0,0, 0.243)	1.31	0.00	(0,0, 0.243)	1.31	0.00
			400	(0, 0, 0.5405)	2.783	(0, 0, 0.5405)	2.783	0.00	(0, 0, 0.5405)	2.783	0.00	(0, 0, 0.5405)	2.783	0.00
160	150	1.313	50	(0,0, 0.0455)	1.521	(0,0, 0.0455)	1.521	0.00	(0,0.56, 0.0723)	0.9864	35.15	(0.944, 1.09, 0.572)	0.573	62.33
			200	(0,0, 0.191)	1.88	(0,0, 0.191)	1.88	0.00	(0,0.516, 0.267)	1.655	11.97	(0.32,0.616 , 0.295)	1.618	13.94
			400	(0,0, 0.413)	2.83	(0,0, 0.413)	2.83	0.00	(0, 0.337, 0.463)	2.766	2.26	(0, 0.337, 0.463)	2.766	2.26

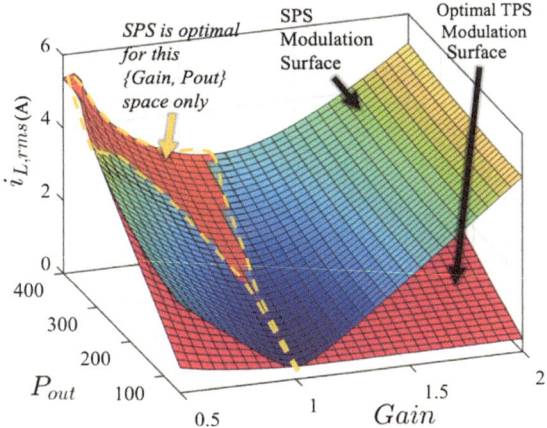

Fig. 3.3 3D mapped $i_{L,rms}$ current comparison between the SPS and optimal TPS modulation schemes for wide output load and gain range

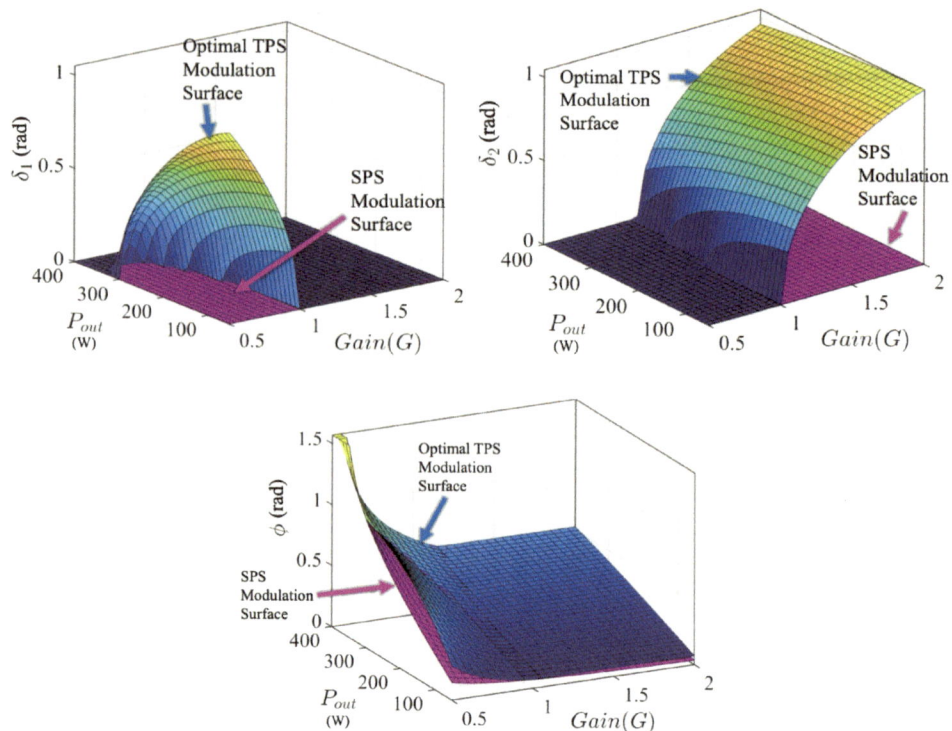

Fig. 3.4 Optimal δ_1, δ_2 and φ under TPS and SPS modulation for variation in output load and gain

3.3   Switching Network Loss-Optimal Modulator Optimization in DC-AC DAB Converter

Figure 3.5 depicts the schematic of the single-phase, single-stage, bidirectional and isolated DAB based dc-ac converter topology, consisting of a DAB dc-dc converter followed by an unfolder and EMI component stage. Following the input dc source V_{dc}, the dc-dc DAB power conversion stage consists of two active bridges coupled by a high-frequency (HF) transformer having a turns ratio of $n_1 : n_2$ and a series connected inductor that can be realized by the leakage inductance of the transformer. The power transfer between the active bridges is controlled by modulating the HF switching node voltages' (v_1 and v_2') duty cycles (δ_1 and δ_2) and their relative phase difference (φ) that results from introducing phase shifts between the gating signals of the active bridge switches, as presented in Fig. 3.2 ($\varphi \in \left[-\frac{\pi}{2}, \frac{\pi}{2}\right]$; $\delta_1, \delta_2 \in \left[0, \frac{\pi}{2}\right]$). Hence, by independently varying the phase shift (φ) and duty cycle (δ_1 and δ_2) control variables, the output voltage of the dc-dc stage $v_{o1}(t)$ is regulated in a shape of a dc voltage that pulsates at twice the ac line frequency f_L (i.e., $v_{o1}(t) = |v_{ac}(t)| = \left|\widehat{V_{ac}}\sin(\omega_L t)\right|$, where $\omega_L = 2\pi f_L$ and $\widehat{V_{ac}}$ is the ac voltage

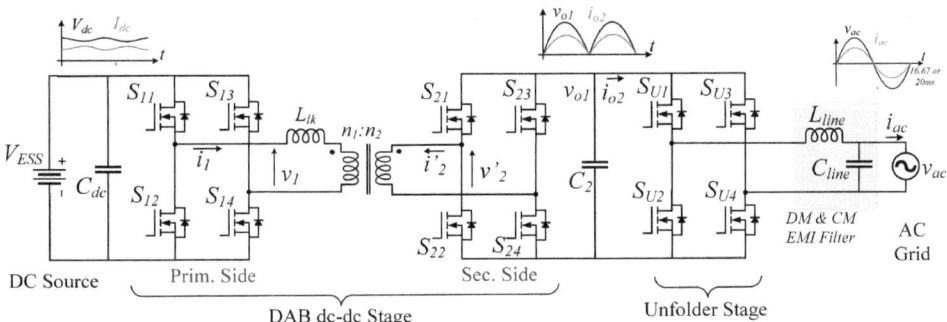

Fig. 3.5 Circuit schematic of dc-ac DAB converter consisting of dc-dc DAB stage followed by an line frequency unfolder stage

peak). The ac link capacitor C_2 is connected at the output stage of the dc-dc DAB to sink in the HF ac current ripple from i_{o1} and the dc component of i_{o1} propagates towards the unfolder stage as i_{o2}.

During converter operation, the state st_U of the unfolder stage switches (S_{U1}–S_{U4}) toggles twice in a time period of ac line voltage $v_{ac}(t)$ such that,

$$st_{U1,U4,\overline{U2},\overline{U3}} = \begin{cases} 1, & \textit{if } v_{ac}(t) > 0 \\ 0, & \textit{if } v_{ac}(t) < 0 \end{cases} \tag{3.4}$$

Following this continuous state change, the pulsating dc voltage $v_{o1}(t)$ is unfolded and forms a sinusoidally varying ac output voltage $v_{ac}(t)$. Note that the voltage drop across the differential mode (DM) output stage EMI filter inductors can be neglected in a steady state. The EMI filter stage is connected at the extreme end of the converter that interfaces the grid and is comprised of DM and common-mode (CM) filter components. All the DM mode filter impedances are equivalently presented as a series connected line inductor L_{line} on the ac line and a capacitor C_{line} connected across the line-neutral terminals.

3.3.1 Loss Function Formulation

We intend to optimize the performance of the dc-ac DAB by obtaining the modulation variables that minimizes the losses in the HF DAB switching network. The losses incurred by the unfolder stage is not dependent on the modulation variables, thus only the losses in the semiconductors of the HF DAB stage are modeled here.

A. *Conduction Loss*:

Total conduction loss in a DAB switching network can be presented as,

$$P_{cond\_loss}\left(\delta_j, \varphi_j, f_{sw}, V_j\right) = \sum_{j=1}^{2} i_{j,RMS}^2 R_j \qquad (3.5)$$

where, $i_{j,RMS}$ is the jth side transformer winding current RMS ($i_{1,RMS}$ is calculated using (3.2) derived using the GHA-enabled modeling approach and $i_{2,RMS} = \frac{n_1}{n_2} i_{1,RMS}$) and $R_j = R_{DS(ON)}$ of the jth bridge switches ($j = 1$ for primary side, $j = 2$ for secondary side).

B. *Switching Loss*:

The switching loss in a DAB converter is determined by the currents flowing through the switches and the voltages they block during transient switching, as well as the operating switching frequency. From the voltage and current expressions of the DAB bridges, presented in Eqs. (2.25), (2.26) and (2.29) of Chap. 2, it can be deduced that all waveforms exhibit half-cycle symmetry: $v_j(t) = -v_j(\pi + t)$ and $i_j(t) = -i_j(\pi + t)$ (for $j = 1$ or 2). Consequently, two fundamental principles regarding the turn-on and turn-off switch currents in any half-bridge leg of a DAB converter are as follows: (a) switches within the same leg of any full bridge experience identical turn-on and turn-off losses due to equivalent currents during these events; (b) the magnitude of the turn-on current for any switch matches exactly with the turn-off current but with opposite polarity. These relationships are crucial for understanding the zero-voltage switching (ZVS) constraints in a TPS-controlled DC-AC DAB, which are further explored and detailed below to model the overall switching losses in the system.

In a DAB converter, the occurrence of ZVS is closely tied to the charging and discharging of the parasitic body capacitance of the switching MOSFET (C_{OSS}), which notably increases as the drain-source voltage (v_{DS}) across the MOSFET decreases. For the H-bridge on the secondary side of a DC-AC DAB converter, the v_{DS} of the MOSFETs can vary widely from 0 V up to the peak AC line voltage (V_{ac}). Due to the higher value of C_{OSS}, achieving ZVS becomes particularly challenging when the output AC line voltage approaches zero and inter-port phase shift nears zero, especially for the switches in the secondary bridge. Therefore, to accurately model switching losses in DAB converters, a regression model-based curve fitting technique can be utilized to represent the C_{OSS} values of the high-frequency (HF) DAB MOSFETs as a closed-form function of v_{DS}, as presented in (3.6). This approach is designed to precisely synthesize the constraints for achieving ZVS and accurately characterize the charging and discharging times of the body capacitors.

$$C_{OSS}(v_{DS}) = \frac{K_1}{\sqrt{v_{DS}}} + K_2 \tag{3.6}$$

The coefficients K_1 and K_2 are determined based on the selection of the power devices and can be calculated from the C_{OSS} versus v_{DS} plots available in the MOSFET datasheets. For example, if GaN MOSFETs EPC2065 and GS-065-030-2-L are used as the primary and secondary side HF switching devices [6, 7], respectively, their nonlinear parasitic body capacitors are modeled as follows.

$$C_{oss,pri} = \left[\frac{3390}{\sqrt{V_{DS}(V)}} + 52 \right] pF;$$

$$C_{oss,sec} = \left[\frac{2790}{\sqrt{V_{DS}(V)}} - 64 \right] pF; \tag{3.7}$$

Figure 3.6 depicts C_{oss} modeling as a function of the V_{DS} voltage. The comparison between the modeled and original C_{oss} data with variation in V_{DS} is portrayed in Fig. 3.6 that highlights good modeling accuracy with a median modeling error of 5%.

Fig. 3.6 Primary and secondary side GaN E-HEMT's C_{oss} modeling as a function of the V_{DS} voltage

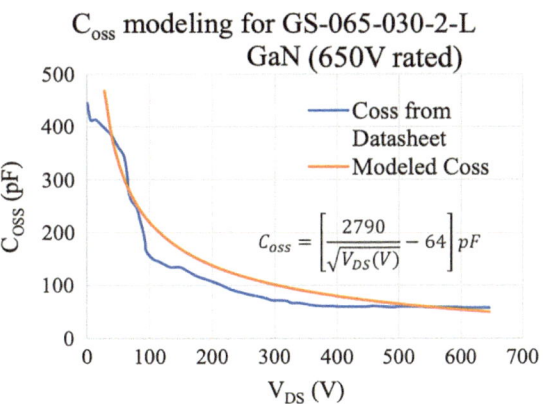

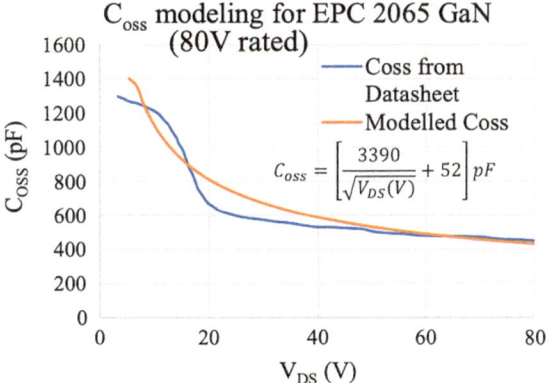

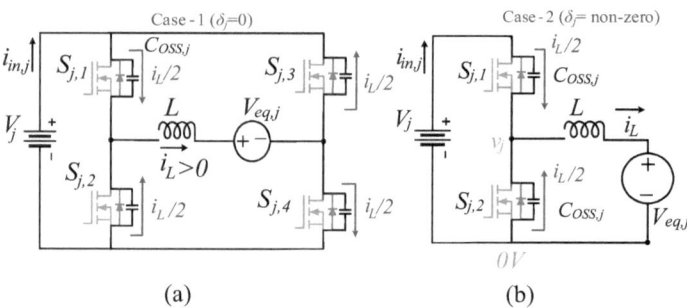

Fig. 3.7 Bridge equivalent DAB circuit models during switching transients

Two different cases of ZVS turn-on event can occur in a TPS controlled dc-ac DAB converter depending on the availability of the bridge voltage duty cycle variable, which are presented in Fig. 3.7. The Case-1 equivalent circuit of the DAB's j-port (where $j = 1$ or 2) applies when there is no duty cycle control involved for that active bridge. Figure 3.7a illustrates the switching transition where $S_{j,1}$ and $S_{j,4}$ are turning off while $S_{j,2}$ and $S_{j,3}$ are preparing to turn on. Here, $v_{eq,j}$ represents the Thevenin equivalent voltage of the output voltage from the other bridge, and i_j denotes the DAB line inductor current referenced to the jth side during this transient event. During this dead-time zone operation, the body capacitors (C_{OSS}) of $S_{j,2}$ and $S_{j,3}$ are charging, while those of $S_{j,1}$ and $S_{j,4}$ are discharging individually due to the current $i_L/2$. The discharging time of the body capacitor of $S_{j,2}$ or $S_{j,3}$, which is the time taken for their drain-source voltage to reach zero, denoted as τ_c, can be calculated using (3.8).

$$\tau_c = \int dt = \frac{2}{i_j} \int_0^{V_j} C_{OSS} dv_{DS,S_{j,1}} = \frac{2\left(K_2 V_j + K_1 \sqrt{V_j}\right)}{i_j} \tag{3.8}$$

In such a circuit configuration, the total energy sunk by the energy sources is,

$$E_{sunk} = \int_0^{\tau_c} \left[v_{eq,j} i_j(\tau) - V_j i_{in,j}(\tau)\right] dt = \int_0^{\tau_c} \left[v_{eq,j}\left(-2C_{OSS,j}\frac{dv_{ds,S_{j,1}}}{dt}\right) - 0\right] dt$$

$$= 2v_{eq,j}\left(K_2 V_j + K_1 \sqrt{V_j}\right) \tag{3.9}$$

where $v_{ds,S_{j,1}}$ is the voltage across the drain-source of $S_{j,1}$. Under this particular switching commutation of case-1, $i_{in,j}(\tau) = 0$ or the inductor current does not leak to the dc link voltage source. However, for case-2 switching scenario, $i_{in,j}(\tau)$ will be equal to $-C_{OSS,j}\frac{dv_{ds,S_{j,1}}}{dt}$. Thus, the expressions for total energy sunk in case-1 and case-2 becomes different.

As the total stored energy in the MOSFET body capacitors does not change during the commutation interval, the condition for ZVS in case-1 can be written as,

$$\frac{1}{2}Li_j(\tau)^2 \geq 2\big(K_2V_j + K_1\sqrt{V_j}\big)v_{eq,j}(\tau) \tag{3.10}$$

The relation in (3.10) suggests that for $v_{eq,j}(\tau) < 0$, ZVS will always be achieved. However, if $v_{eq,j}(\tau)$ is positive, the minimum required port output current to attain ZVS can be expressed as, $|i_j| \geq 2\sqrt{\dfrac{\big(K_2V_j + K_1\sqrt{V_j}\big)}{L}}\,v_{eq,j}$. Another condition of the ZVS under this scenario needs the dead-time (t_d) to be larger than the drain-source voltage discharging time, or $t_d > \tau_c$.

All the possible cases of DAB switching transients and their respective ZVS requirements are identified and presented in Table 3.3 as a summary.

Upon successfully deriving the ZVS conditions for a pulsating dc output DAB operation, the overall switching loss in the active devices need to be quantified. While all the switches undergo turn-off switching losses due to non-zero turn-off switch current and drain-source voltage overlap, the turn-on loss due to the switch voltage-current overlap can be eliminated when a hard-switching event is avoided. Thus, the hard-switching event informed switching loss due to voltage-current overlap in the DAB stage under study can be presented as,

Table 3.3 Switching current informed ZVS constraints for jth active bridge (j=1 or 2; $\varphi_1 = 0$)

Case	Turn-on switches	ZVS condition	Switching time (τ)
$\begin{cases} \delta_j = 0 \\ i_j(\tau) > 0 \end{cases}$	S_{j1}, S_{j4}	$\frac{1}{2}Li_j(\tau)^2 \geq 2(K_2V_j + K_1\sqrt{V_j})v_{eq,j}(\tau)$	$\varphi_j + \pi$
$\begin{cases} \delta_j = 0 \\ i_j(\tau) < 0 \end{cases}$	S_{j2}, S_{j3}	$\frac{1}{2}Li_j(\tau)^2 \geq -2(K_2V_j + K_1\sqrt{V_j})v_{eq,j}(\tau)$	φ_j
$\begin{cases} \delta_j \neq 0 \\ i_j(\tau) > 0 \end{cases}$	S_{j2}	$\frac{1}{2}Li_j(\tau)^2 \geq (2v_{eq,j}(\tau)V_j - V_j^2)(K_1 + \frac{K_2}{\sqrt{V_j}})$	$\varphi_j + \pi - \delta_j$
$\begin{cases} \delta_j \neq 0 \\ i_j(\tau) < 0 \end{cases}$	S_{j1}	$\frac{1}{2}Li_j(\tau)^2 \geq (-2v_{eq,j}(\tau)V_j - V_j^2)(K_1 + \frac{K_2}{\sqrt{V_j}})$	$\varphi_j - \delta_j$
$\begin{cases} \delta_j \neq 0 \\ i_j(\tau) > 0 \end{cases}$	S_{j3}	$\frac{1}{2}Li_j(\tau)^2 \geq (2v_{eq,j}(\tau)V_j + V_j^2)(K_1 + \frac{K_2}{\sqrt{V_j}})$	$\varphi_j + \pi + \delta_j$
$\begin{cases} \delta_j \neq 0 \\ i_j(\tau) < 0 \end{cases}$	S_{j4}	$\frac{1}{2}Li_j(\tau)^2 \geq (-2v_{eq,j}(\tau)V_j + V_j^2)(K_1 + \frac{K_2}{\sqrt{V_j}})$	$\varphi_j + \delta_j$

$$P_{sw,v-i}\left(\delta_j,\varphi_j,f_{sw},V_j\right)=$$

$$\begin{cases} 2\cdot\displaystyle\sum_{j=1}^{2}\left[\begin{array}{l}2V_j\left|i_j(\varphi_j)\right|\left|f_{sw}t_{off,j}+\\2V_j\left|i_j(\varphi_j)\right|\left|f_{sw}t_{on,j}\right.\end{array}\right]; \delta_j=0 \\[24pt] \displaystyle\sum_{j=1}^{2}\sum_{k=1}^{2}\left[\begin{array}{l}2V_j\left|i_j(\tau_{j,k})\right|\left|f_{sw}t_{off,j,k}+\\2V_j\left|i_j(\tau_{j,k})\right|\left|f_{sw}t_{on,j,k}\right.\end{array}\right]; \delta_j>0 \end{cases} \tag{3.11}$$

ZVS-critical current requirements for soft turn-on events are described by the conditions placed in Table 3.3. When the instantaneous tank current polarity as well as the deadtime duration are favorable to discharge FET body capacitor but the tank current magnitude is not critically enough to turn on the FET with ZVS, the current will hit zero right when the drain-source voltage reaches its valley. Rising edge triggering of the complementary FET gate-source right at the valley will cause zero-current-switching (ZCS) turn-on and hence zero V-I product. The dominant loss in such as a case would be CV^2 loss based on the residue voltage on the FET body capacitor. Switching at a voltage other than valley would result in non-zero capacitor current and non-ZCS turn-on that would need to account for the CV^2 loss as well as the V-I product-based loss as per (3.11). In case of a tank current with over ZVS-critical value but insufficient deadtime, there is a non-zero source-to-drain current at the rising edge instant of gate-source voltage and incomplete discharge of FET capacitor, so there will be V-I product loss as well as CV^2 loss based on the residue drain-source voltage, as expressed in (3.12). If the tank current polarity is not favorable for ZVS turn-on, there will be full overlap between voltage fall and current rise. The turn-off current, turn-off and turn-on time of kth (k = 1, 2) half-bridge of the jth bridge are highlighted as $\left|i_{j,k}(\tau_{j,k})\right|$, $t_{off,j,k}$ and $t_{on,j,k}$, respectively. Here, the t_{off} and t_{on} times of the MOSFET under use depend on the MOSFET's characteristics such as its parasitic capacitance values i.e., gate to drain capacitance C_{gd}, input capacitance C_{in}, on-state drain-source resistance R_{on}, gate plateau voltage V_{pl}, gate voltage threshold V_{th}, as well as external circuit conditions such as gate resistance R_g, drain current during on-state I_D, off-state blocking voltage of the MOSFET V_{DC}. The parameters dependent on the MOSFET's device characteristics such as C_{gd}, C_{in}, V_{th} and V_{pl} can be found from the datasheet of the corresponding MOSFET. The turn-off switching time of any MOSFET can be written as, $t_{off}=t_{fi}+t_{rv}$, where t_{fi} represents the time required for the fall of drain current (I_D) to zero and is denoted as $t_{fi}=R_gC_{in}\ln\left(\frac{V_{pl}}{V_{th}}\right)$ and t_{rv} represents the time required for the rise of drain-source voltage (V_{DS}) to dc link voltage (V_{DC}) and is denoted as $t_{rv}=\frac{R_gC_{gd}(V_{DC}-I_DR_{on})}{V_{pl}}$. Similarly, the turn-on switching time under hard-switching event can be obtained as, $t_{on}=R_gC_{in}\ln\left(\frac{V_{Dr}-V_{th}}{V_{Dr}-V_{pl}}\right)+\frac{C_{gd}(V_{DC}-I_DR_{on})R_g}{V_{Dr}-V_{pl}}$, where R_g denotes the total gate resistance, R_{on} denotes the on-state resistance of the MOSFET, C_{in} denotes the MOSFET input capacitance, V_{Dr} represents the gate drive supply voltage, V_{pl} denotes the plateau voltage, and V_{th} represents the MOSFET threshold voltage.

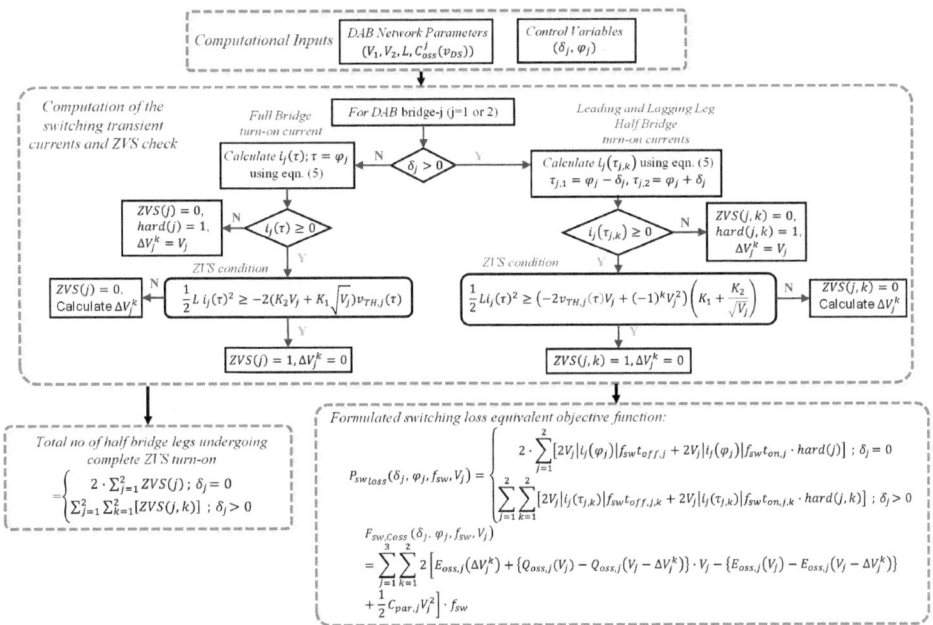

Fig. 3.8 Flowchart depicting the switching loss computation in a DAB dc-dc stage

For better understanding, the switching loss calculation procedure for the DAB dc-dc stage under a specific input/output voltage condition is clearly described in the form of a flowchart in Fig. 3.8.

Apart from the losses incurred by overlap between switching voltage and current during any switching event, the incomplete charge and discharge of the MOSFET's body capacitors (C_{oss}) also add on to the switching loss under and partial or full hard-switching event. From the discussion of energy dissipation during incomplete ZVS and hard-switching events, delineated in Chap. 2, the C_{oss} related loss in a TAB converter can be estimated as,

$$F_{sw,Coss}(\delta_j, \varphi_j, f_{sw}, V_j) = \sum_{j=1}^{3}\sum_{k=1}^{2} 2\Big[\big[E_{oss,j}\big(\Delta V_j^k\big)\big[+\big\{Q_{oss,j}(V_j) - Q_{oss,j}\big(V_j - \Delta V_j^k\big)\big\}$$

$$\Delta V_j - \big\{E_{oss,j}(V_j) - E_{oss,j}\big(V_j - \Delta V_j^k\big)\big\} + \frac{1}{2}C_{par,j}V_j^2\Big] \cdot f_{sw}. \tag{3.12}$$

Here, $E_{oss,j}\big(\Delta V_j^k\big)$ denotes the energy stored in the nonlinear capacitance C_{oss} of the MOSFET connected at k-th switching leg of j-th bridge due to the presence of ΔV_j^k voltage across it during its turn-on instant. This remaining drain-source voltage ΔV_j^k at turn-on instant can be found out by solving the energy balance as shown in Chap. 2 (Ref. (2.72)). For any drain-source voltage V_{DS}, the stored energy at C_{oss} is $E_{oss}(V_{DS}) = \int_0^{V_{DS}} C_{oss}(v) \cdot v dv = \frac{1}{2} C_E(V_{DS}) \cdot V_{DS}^2$ and can be calculated from the $C_{oss} - V_{DS}$ plot given in the device datasheet. Similarly, $Q_{oss}(V_{DS})$ denotes the amount of charge stored in C_{oss} due to V_{DS} voltage present across it and can be calculated as, $Q_{oss}(V_{DS}) = \int_0^{V_{DS}} C_{oss}(v) \cdot dv = C_Q(V_{DS}) \cdot V_{DS}$. The C_{par} is the parasitic switching node capacitance that appears due to the printed circuit board design, parasitic capacitances of the transformer etc. and needs to be charged and discharged everytime during any switching commutation. In case of complete soft-switching transition, ΔV_j^k will be 0, while, for a complete hard-switching, $\Delta V_j^k \rightarrow V_j$.

Finally, a cost function $F_{sw\_loss}(\delta_j, \varphi_j, f_{sw})$ that accurately models the loss in a TAB switching network due to the switching action can be derived as a sum of $F_{sw\,loss}(\delta_j, \varphi_j, f_{sw})$ and $F_{Coss}(\delta_j, \varphi_j, f_{sw})$.

$$F_{sw\_loss\_total}(\delta_j, \varphi_j, f_{sw}, V_j) = F_{sw,v-i}(\delta_j, \varphi_j, f_{sw}, V_j) + F_{sw,Coss}(\delta_j, \varphi_j, f_{sw}, V_j) \quad (3.13)$$

C. MOSFET Reverse Conduction Loss:

The third major loss occurs in the DAB switching network due to the reverse conduction of the MOSFETs through their intrinsic body-diodes during the deadtime between the same leg switch gate pulses after the switching commutation interval ends. This loss can be expressed as,

$$P_{rev\_cond}(\delta_j, \varphi_j, f_{sw}, V_j) = \sum_{j=1}^{2} \sum_{k=1}^{2} |i_j(\tau_{j,k})| V_{F_{j,k}} f_{sw}(t_{d_{j,k}} - \tau_{c_{j,k}}) \quad (3.14)$$

where $|i_j(\tau_{j,k})|$ and $V_{F_{j,k}}$ are the switching turn-on current and the forward voltage drop of the body diodes of the MOSFETs connected at kth (k = 1, 2) half-bridge of the jth H-bridge. Here, $t_{d_{j,k}}$ represents the imposed dead time between the high side and low side devices of the kth leg and $\tau_{c_{j,k}}$ is the switching commutation time taken during charging and discharging the drain-source body capacitor of the FET. To reduce the reverse conduction loss and to ensure ZVS turn-on of the switches, an effort is made in this study to optimally choose the deadtime between the gate pulses of the switching devices connected in the same leg of any active H-bridge cell. An adequate amount of dead time needs to be present between the high-side and the low side switch of a switching leg in a DAB converter in order to avoid any shoot-through fault as well as to facilitate the ZVS turn-on event even under lower switching node dv/dt. However, a higher amount of dead time incurs higher conduction loss in the switch. As observed from (3.14), to restrict the reverse diode conduction losses caused by the applied extended dead-time, the optimal value of the deadtime must be chosen as slightly above the maximum possible value of the τ_c. Therefore, the optimal choice of deadtime is crucial in an efficiency critical DAB application.

D. *Total Loss Function Formulation*:

The target is set to minimize the accurately formulated above-mentioned losses associated with the dc-dc DAB stage switching network over the output ac line cycle of the dc-ac converter by optimally varying the TPS modulation variables (δ_1, δ_2 and φ) and the switching frequency (f_{sw}) of dc-dc stage.

The objective or the cost function for the optimization algorithm can be defined in various ways targeting different loss elements in the HF DAB stage such as conduction, switching loss or ZVS extension, according to the user need. However, in the proposed method, the cost function is modeled as the total switching network loss in the DAB dc-dc stage to maximize the system efficiency, as given in (3.15).

$$f_{cost}(x) = P_{cond\_loss} + P_{sw\_loss} + P_{rev\_cond} \tag{3.15}$$

Here, P_{cond_loss}, P_{sw_loss}, and P_{rev_cond} are the conduction, switching and reverse conduction losses in switching devices for a particular voltage conversion.

3.3.2 Optimization Constraints Formulation Based on Power Flow and Soft-Switching Operation

There are constraints imposed on all control variables and design variables. While design constraints are typically imposed on the system passives such as line inductance, capacitance, and transformer turns ratios based on efficiency and power density targets, the control variable constraints are applied to the optimization routine framework that is mostly implemented in a digital control platform. The constraints pertain to the extreme values attained by the phase angle, duty ratio, and switching frequency. While there are some hard constraints set by theoretical bounds, the other constraints could be flexible and varying across different designs. For example, the switching frequency bound can be imposed by maximum allowable turn-off losses that directly relate to the efficiency profile or by EMI filter sizing that impacts the system power density or by program execution feasibility offered by the digital control platform employed.

These constraints can be divided into two categories:

(1) Non-linear Constraints: The power demand by the load sets this constraint where a desired instantaneous power needs to be supplied to the ac grid during any particular ac cycle phase angle for a dc-ac DAB. Therefore, this nonlinear equality constraint can be rewritten as,

$$c_{eq}(x) = 0 \text{ or, } P_o(x) - P_{out}(t) = 0 \tag{3.16}$$

where, $P_o(x)$ can be derived from (3.15) and $P_{out}(t)$ is the instantaneous load power.

$$P_o = \frac{4}{\pi^3 f_{sw}} \cdot \sum_{k=1}^{2m+1} \frac{1}{k^3} \left[\frac{V_1 V_2}{L} \cos(k\delta_2) \cos(k\delta_1) \sin(k\varphi) \right] \tag{3.17}$$

(2) Functions imposing limitation on the decision variables: These types of functions (3.18)–(3.20) set the upper and lower bounds to the x so that the solution stays inside the feasible zone.

$$0 \le \delta_1 \le \frac{\pi}{2} \ (3.18); \quad 0 \le \delta_2 \le \frac{\pi}{2} \ (3.19); \quad -\frac{\pi}{2} \le \varphi \le \frac{\pi}{2} \ (3.20);$$

$$f_{sw,min} \le f_{sw} \le f_{sw,max} \tag{3.21}$$

As for the switching frequency selection, depending on the use case, the efficiency is often prioritized while power density does not play as a high-priority performance metric, especially for stationary storage power conversion application. In such designs, the frequency is kept reasonably low (in 50–70 kHz range) not only to minimize the switching losses, but also to push the fundamental switching component away from conducted EMI spectrum zone, so the EMI filter is designed with respect to a higher order harmonic, reducing the filter volume.

Since the dc-ac DAB requires a gain and load variation from 0 to $v_{ac,pk}$ and 0 to P_{pk}, respectively over each half ac line period, for a fixed f_{sw} operation, TPS controlled DAB converter needs an optimally chosen value of inductor L that ensures the minimal conduction and switching losses weighted across different gain range. Nevertheless, such a variable L-based DAB design is practically very complex. Therefore, instead of varying L, f_{sw} can be varied equivalently in a range to ensure the efficient dc-ac DAB operation. In a variable frequency-operated dc-ac DAB, the upper limit of the DAB switching frequency f_{sw} can be selected as 150 kHz or lower to facilitate a compact passive component sizing without leading to excessive frequency related losses such as switching losses and transformer winding losses. Once a magnetic core geometry along with absolute turns count is selected for fabricating a transformer based on the power density requirement, the choice of lowest switching frequency determines the highest flux density and core loss. Thus, the lower bound of f_{sw} is determined based on the DAB transformer core saturation criteria.

Furthermore, to even improve the $f_{sw,min}$ choice, a second order constraint is imposed on the optimization algorithm based on the maximum allowable computed total harmonic distortion (THD) on the v_{ac}, i.e., 2%. In this process, THD in the output voltage is calculated after the optimization algorithm completes searching for the optimal decision variable set (x) for a complete ac line cycle. If the analytically calculated THD is above the limit, the choice of $f_{sw,min}$ is decreased by 5 kHz in each iteration. The optimization algorithm flowchart is depicted in Fig. 3.9.

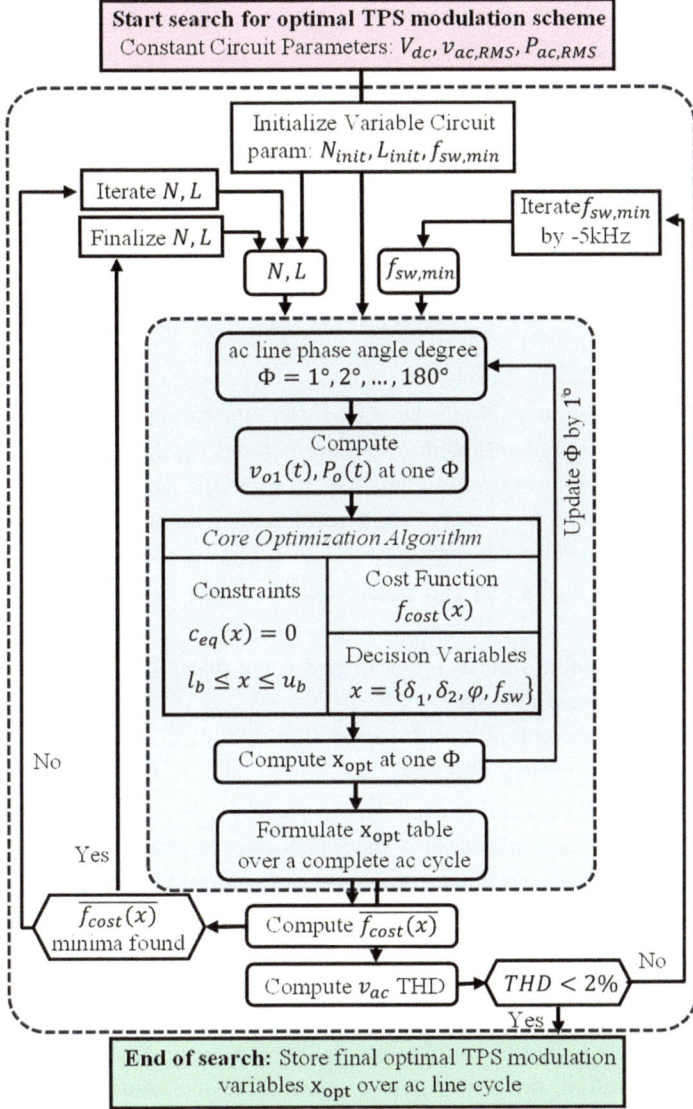

Fig. 3.9 Detailed flowchart for determining the proposed modulation scheme for optimally designed dc-ac DAB converter using multi-variable non-linear constrained optimization algorithm

3.3.3   Co-optimization of DAB Circuit Parameter Variables

With the aim of improving the physical layout of the DAB switching network and suggesting an optimal TPS modulation scheme to minimize losses, the optimization process considers two key circuit parameters to be chosen as design variables: the transformer turns ratio N and the HF line inductance L. Choosing these two parameters is straightforward for a constant input–output dc-dc DAB converter. However, due to the pulsating nature of the output V_{o1}, selecting the optimal values for N and L becomes challenging. The upper limit on the value of L is determined from the maximum power flow criteria, as given in (3.22) (maximum power flow in a DAB is observed when $\delta_1 = 0, \delta_2 = 0, \varphi = \frac{\pi}{2}$).

$$P_{o,max} \leq \frac{4}{\pi^3 f_{sw,max} L} \cdot \sum_{k=1}^{2m+1} \frac{1}{k^3}\left[V_1 V_2 \sin\left(\frac{k\pi}{2}\right)\right] \tag{3.22}$$

While the upper limit of N can be set to $\frac{\widehat{V_{ac}}}{V_{dc}}$, the target is to choose optimal N, L pair that produces the most efficient switching network design. Therefore, the proposed switching modulation optimization algorithm is solved in an iterative way generating a set of x_{opt} for different N, L pair design combination as an input. From the resultant x_{opt} set over a line cycle, the average total loss $\overline{f_{cost}(x)}$ over a line cycle ($\overline{f_{cost}(x)} = \frac{1}{2\pi f_L} \int_0^{2\pi f_L} f_{cost}(x))$ is computed and finally one N, L pair is selected that results in the least $\overline{f_{cost}(x)}$.

As a case study, this exercise is performed for a dc-ac DAB converter with its key specifications shown in Table 3.4.

The result from the proposed N, L pair design optimization technique is presented in Fig. 3.10, where it is clearly observed that $N = 3$ and $L_2 = 20\mu H$ generates the least inductor current RMS over a line cycle ($i_{1,lineRMS} = \sqrt{\frac{1}{2\pi f_L} \int_0^{2\pi f_L} i_{1,RMS}^2(x)}$), i.e., 17.1 A. Thus, during the final converter design, the transformer is designed in such a way that has a turns ratio of 1:3 and the secondary side power transfer inductance is set at 20 μH. Now, this optimization could be looked from another perspective if the design intention was to maximize the power density, in which case, the designer could have looked for options to integrate the secondary power transfer inductance as part of the controllable leakage of the transformer. To produce a 20 μH leakage inductance as part of a transformer with 1:3 turns ratio, the design would require too high count of secondary turns alongside a larger core volume. Therefore, the power density optimal design could have led to a different (N, L) set that might not have been the most efficient.

Table 3.4   Fabricated dc-ac DAB converter circuit components

Circuit parameters	Values
Input DC voltage (V_1)	30–50 VDC
Primary H-bridge switches	EPC2065 100 V 60 A 3.6 mΩ GaN FET
Primary side DC link capacitor (C_1)	9.9 mF/100 V
AC output voltage (v_{ac})	120–230 Vac
Rated output power (P_O)	500 W
Secondary HF H-bridge switches	GS-065-030-2-L 650 V 30 A 50 mΩ GaN FET
Secondary side pulsating DC link capacitor (C_2)	3.29 µF/500 V
Secondary side unfolder bridge switches	C3M0120065D 650 V 22 A SiC MOSFET
Planar transformer core	FR45810EC E shape
Transformer turns ratio (N_1:N_2)	4:12
Magnetizing inductance (L_m)	290 µH
Total leakage inductances (L_1, L_2')	0.54 and 21 µH
DM Filter Inductance (L_{line})	80 µH
DM Filter Capacitance (C_{line})	2 µF/310 Vac
Switching frequency (f_{sw})	50–150 kHz

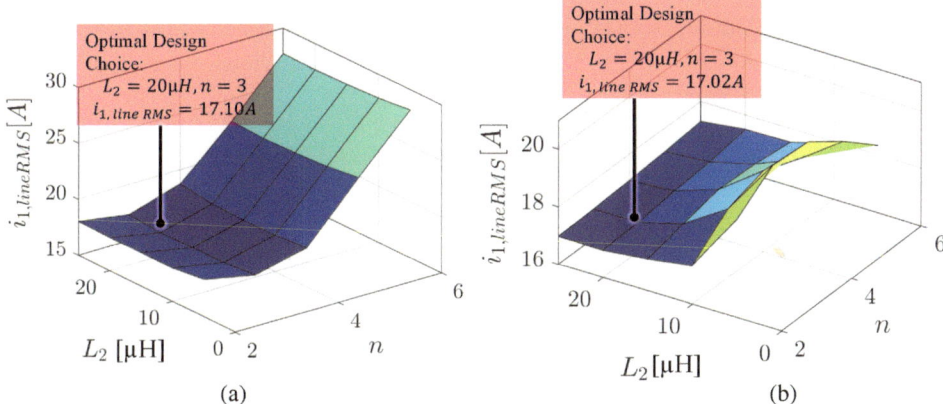

Fig. 3.10   Calculated $i_{1,lineRMS}$ for a widely varying (N, L_2) combination for **a** 40 V_{dc} to 120 V_{ac} and **b** 40 V_{dc} to 230 V_{ac} conversion at 500 W ac load

3.4　Analytical Formulation of THD Model of a DC-AC DAB Converter

Often, in design optimization problems for dc-ac DAB converters, THD is kept as an objective function or as an inequality constraint for the system to be complying with the IEEE standards for grid integration. Therefore, to solve for the modulator optimization and converter control, THD needs to be formulated as a closed form mathematical expression as a function of the control degrees of freedom, such as frequency, phase, and duty variables. This section describes an analytical method for THD computation in any frequency or phase or duty-controlled dc-ac converters.

THD in the output ac voltage or current appears due to the presence of the higher order switching harmonic components in the secondary bridge output current $i_{o1}(t)$ according to Fig. 3.11. Therefore, an approach to determine the amplitudes of all the harmonics including the fundamental component in $i_{o1}(t)$ is to decompose $i_{o1}(t)$ into its Fourier components using Fast Fourier Transformation (FFT). The bridge output current that flows into the ac link capacitor C_2 can be related with the previously formulated DAB inductor current $i_1(t)$ based on the secondary side bridge duty cycle (δ_2) and phase-shift (φ), as given in (3.23).

$$i_{o1}(t) = \begin{cases} 0, (\varphi - \delta_2) < \omega t < (\varphi + \delta_2) \\ \frac{i_1(t)}{N}, (\varphi + \delta_2) < \omega t < (\pi + \varphi - \delta_2) \\ 0, (\pi + \varphi - \delta_2) < \omega t < (\pi + \varphi + \delta_2) \\ -\frac{i_1(t)}{N}, (\pi + \varphi + \delta_2) < \omega t < (2\pi + \varphi - \delta_2) \end{cases} \quad (3.23)$$

$i_{o1}(t)$, which is the rectified output current, is a periodic signal with twice switching frequency ($2f_{sw}$) component due to bridge rectification. The nth harmonic coefficient C_n for $i_{o1}(t)$ decomposition can be deduced as:

$$C_n = \frac{\omega}{\pi} \int_0^{\frac{\pi}{\omega}} i_{o1}(\tau) \cdot e^{-i2n\omega t} dt = \frac{\omega}{\pi} \int_0^{\frac{2\delta_2}{\omega}} 0 \cdot e^{-i2n\omega t} dt + \frac{\omega}{\pi} \int_{\frac{2\delta_2}{\omega}}^{\frac{\pi}{\omega}} \frac{i_1(t)}{N} \cdot e^{-i2n\omega t} dt \quad (3.24)$$

Fig. 3.11 Equivalent Circuit representation of the dc-ac microinverter output stage

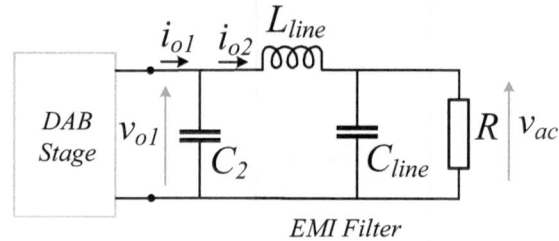

EMI Filter

where, $\tau = t - \frac{\varphi - \delta_2}{\omega}$. Now, applying the time shift property of Fourier series, the nth harmonic coefficient D_n of the actual bridge output current $i_{o1}(t)$ is shown in (3.25).

$$D_n = C_n \cdot e^{-i2n(\varphi - \delta_2)} \tag{3.25}$$

In order to solve (3.24), the DAB inductor current $i_1(t)$ of (2.29) (see Chap. 2) is presented in its complex Euler form,

$$i_1(t) = \sum_{k=1}^{2m+1} \left[\frac{A_k}{2i} \left(e^{ik\omega t} - e^{-ik\omega t} \right) + \frac{B_k}{2} \left(e^{ik\omega t} + e^{-ik\omega t} \right) \right] \tag{3.26}$$

Furthermore, putting (3.24) and (3.26) in (3.25), the complex D_n coefficient is synthesized and is given in (3.27).

$$D_n = X_{n,k} + jY_{n,k} \tag{3.27}$$

Here, the real and imaginary parts of the D_n are presented as $X_{n,k}$ and $Y_{n,k}$, respectively and are expanded in (3.28).

$$X_{n,k} = \frac{1}{\pi} \sum_{k=1}^{2m+1} \begin{bmatrix} \frac{A_k}{2(k-2n)}[-\cos\{\pi k - 2n(\pi + \varphi - \delta_2)\} + \cos\{2k\delta_2 - 2n(\varphi - \delta_2)\}] \\ +\frac{B_k}{2(k-2n)}[\sin\{\pi k - 2n(\pi + \varphi - \delta_2)\} - \sin\{2k\delta_2 - 2n(\varphi - \delta_2)\}] \\ +\frac{A_k}{2(k+2n)}[-\cos\{\pi k + 2n(\pi + \varphi - \delta_2)\} + \cos\{2k\delta_2 + 2n(\varphi + \delta_2)\}] \\ +\frac{B_k}{2(k+2n)}[\sin\{\pi k + 2n(\pi + \varphi - \delta_2)\} - \sin\{2k\delta_2 + 2n(\varphi + \delta_2)\}] \end{bmatrix};$$

$$X_{n,k} = \frac{1}{\pi} \sum_{k=1}^{2m+1} \begin{bmatrix} \frac{A_k}{2(k-2n)}[-\cos\{\pi k - 2n(\pi + \varphi - \delta_2)\} + \cos\{2k\delta_2 - 2n(\varphi - \delta_2)\}] \\ +\frac{B_k}{2(k-2n)}[\sin\{\pi k - 2n(\pi + \varphi - \delta_2)\} - \sin\{2k\delta_2 - 2n(\varphi - \delta_2)\}] \\ +\frac{A_k}{2(k+2n)}[-\cos\{\pi k + 2n(\pi + \varphi - \delta_2)\} + \cos\{2k\delta_2 + 2n(\varphi + \delta_2)\}] \\ +\frac{B_k}{2(k+2n)}[\sin\{\pi k + 2n(\pi + \varphi - \delta_2)\} - \sin\{2k\delta_2 + 2n(\varphi + \delta_2)\}] \end{bmatrix};$$

$$Y_{n,k} = \frac{1}{\pi} \sum_{k=1}^{2m+1} \begin{bmatrix} \frac{A_k}{2(k-2n)}[-\sin\{\pi k - 2n(\pi + \varphi - \delta_2)\} + \sin\{2k\delta_2 - 2n(\varphi - \delta_2)\}] \\ +\frac{B_k}{2(k-2n)}[-\cos\{\pi k - 2n(\pi + \varphi - \delta_2)\} + \cos\{2k\delta_2 - 2n(\varphi - \delta_2)\}] \\ +\frac{A_k}{2(k+2n)}[\sin\{\pi k + 2n(\pi + \varphi - \delta_2)\} - \sin\{2k\delta_2 + 2n(\varphi + \delta_2)\}] \\ +\frac{B_k}{2(k+2n)}[\cos\{\pi k + 2n(\pi + \varphi - \delta_2)\} - \cos\{2k\delta_2 + 2n(\varphi + \delta_2)\}] \end{bmatrix}$$

$$\tag{3.28}$$

In steady state, during a switching cycle the DAB circuit parameters (V_{dc}, v_{o1}) and its TPS control parameter dependent terms in (3.28), such as $A_k, B_k, \varphi, \delta_1$ and δ_2 remain constant. Thus, in a switching cycle ($1/f_{sw}$), the amplitude of the nth harmonic current component in $i_{o1}(t)$ can be determined by (3.29).

$$|D_n| = \sqrt{|X_{n,k}|^2 + |Y_{n,k}|^2} \tag{3.29}$$

Furthermore, to determine the complete FFT analysis of $i_{o1}(t)$ over an ac line cycle, the $|D_n|$ is computed for each of the $1°$ phase angle interval across the ac line cycle of $0°$–$360°$, where the system is considered in steady state with constant circuit parameters and static operating condition. Hence, the final RMS amplitude of the nth harmonic current component in $i_{o1}(t)$ over an ac line cycle is deduced by (3.30).

$$|A_n| = \sqrt{\frac{\sum_{p=1,2,...}^{180} |D_{n,p}|^2}{180}} \tag{3.30}$$

To complete the FFT analysis of the output ac voltage $v_{ac}(t)$ and analytically determine its THD, we evaluate the v_{ac}/i_{o1} transfer function from the equivalent circuit of the dc-ac converter output stage, presented in Fig. 3.10, which can be expressed as,

$$\frac{v_{ac}(s)}{i_{o1}(s)} = \frac{R}{1 + sR(C_2 + C_{line}) + s^2 L_{line} C_2 + s^3 L_{line} R C_2 C_{line}} \tag{3.31}$$

Here, R is the equivalent ac load of the converter, i.e., $R = \frac{P_o^2}{v_{ac,RMS}}$.

Now, the nth harmonic voltage amplitude ($|B_n|$) in $v_{ac}(t)$ can be formulated using (3.32),

$$|B_n| = |A_n| \cdot \left| \frac{v_{ac}(j2\pi nf_L)}{i_{o1}(j2\pi nf_L)} \right| \tag{3.32}$$

Lastly, the THD in $v_{ac}(t)$ can be attained as,

$$THD = \sqrt{\frac{\sum_{n=2,3,4,....} B_n^2}{B_1^2}} \tag{3.33}$$

3.5   Results of the Efficiency-Optimal DC-AC DAB Modulator and Its Comparison with State-of-the-Art Modulators

Based on the gathered information on the optimal control variable dataset from the optimized modulator, a quantitative as well as qualitive analysis of the dc-ac DAB results are carried out in this section. To verify the usefulness and advantages of the single-stage variable frequency TPS DAB dc-ac converter over the constant frequency controlled SPS and TPS DAB modulation scheme, a comparative study is presented. The fixed-frequency TPS modulation is a direct derivation from TPS modulation framework, as shown in Fig. 3.12 and is designed for 100 kHz switching frequency. While we could have selected any value other than 100 kHz, the fundamental differences between a variable f_{sw} TPS and a constant f_{sw} TPS would still have prevailed.

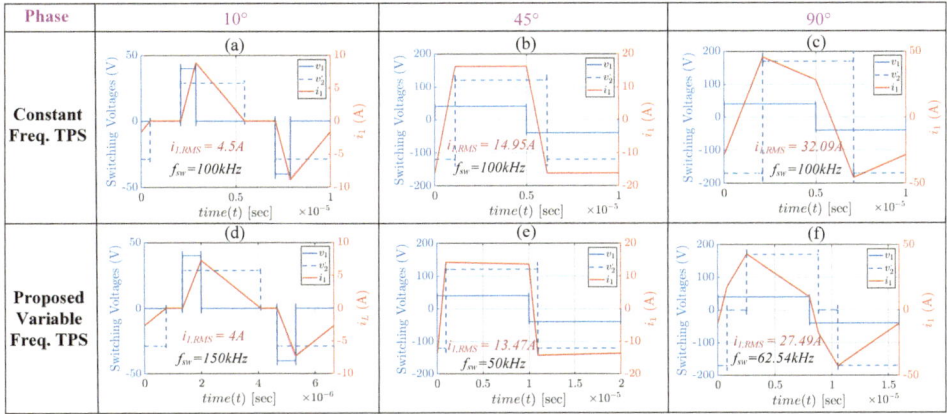

Fig. 3.12 Steady state zoomed-in switching waveform of the dc-ac DAB converter comparing a variable frequency controlled TPS and constant frequency controlled TPS at $\Phi = 10°, 45°$ and $90°$ while supporting a 500 W ac load at 120 V_{RMS} with a 40 Vdc input source

Using the converter parameters outlined in Table 3.1, a steady-state analysis is performed and the respective waveforms from the dc-ac converter under the proposed optimal modulation and constant frequency TPS modulation scheme at three different ac line phase angle Φ are presented in Fig. 3.12 for 40 Vdc to 120 Vac conversion at 500 W average power. Figure 3.12a, d portray the switching waveforms at $\Phi = 10°(v_{o1}(t) = 29.47\,V, P_o(t) = 30.15\,W)$, where the DAB inductor current i_1 achieves RMS value of 4 A and 4.5 A under the proposed and existing DAB modulation strategy, respectively. The proposed modulation attains a f_{sw} of 149 kHz (i.e., the upper bound) during this operation. Figure 3.11(b) and (e) display the steady state DAB switching waveforms at $\Phi = 45°(v_{o1}(t) = 120\,V, P_o(t) = 500\,W)$. During this unity gain ($\frac{v_{o1}}{NV_{dc}} = 1$) voltage transfer mode (Fig. 3.12f), the proposed optimizer selects the optimal f_{sw} to be 50 kHz and with the corresponding optimal phase-shift values, $i_{1,RMS}$ attains a lower value of 13.47. A compared to the constant frequency based control (14.95 A) at 100 kHz. Moreover, during the transfer of maximum power ($P_o(t) = 1000\,W$) to the AC load, specifically at $\Phi = 90°$, the preferred DAB modulation method adjusts the switching frequency (f_{sw}) to 62.5 kHz. This adjustment achieves an RMS current in the HF inductor of 27.49 A, which is 14.3% lower than that achieved with the 100 kHz TPS DAB operation. These lower amount of $i_{1,RMS}$ helps to attain a lesser conduction loss in the MOSFETs and the transformer winding. It can also be observed from Fig. 3.12 that the peak inductor current is reduced with application of the proposed DAB switching technique resulting in lower switching loss under any Φ operation.

In order to gain a better understanding of the optimal dc-ac DAB modulator's benefit over the constant frequency controlled SPS and optimal TPS strategy, different metrics including DAB control variables, $i_{1,RMS}$, turn-off currents of the DAB switching legs, number of HF legs undergoing ZVS, and HF switching network losses are observed over a half ac line cycle (0 to π) for the 40 Vdc to 120 Vac at 500 W power conversion mode and are presented as graphs in Fig. 3.13. The key take-aways from the comparison are:

(a) With this optimal modulator, the f_{sw} varies between 50 and 149 kHz range depending on the instantaneous output power and the input–output voltage gain m condition. At lighter load and lower m, f_{sw} stays at 149 kHz; near the unity m, it stays at 50 kHz and near the peak v_{ac} voltage when maximum power is transferred, f_{sw} stays near 62.5 kHz. Due to this, during peak power flow, the optimal DAB modulator requires less phase shift over the traditional TPS DAB that is switching at 100 kHz, resulting in a considerably lower $i_{1,RMS}$ near $\Phi = 90°$. This leads to a saving of 8 W conduction loss at 1000 W instantaneous output power compared to 100 kHz TPS DAB control scheme. Further, while the optimal modulator has attained a DAB inductor current RMS $i_{1,lineRMS}$ of 17.1 A over a complete 60 Hz line cycle, the 100 kHz switched DAB observes a $i_{1,lineRMS}$ of 19.02 A.

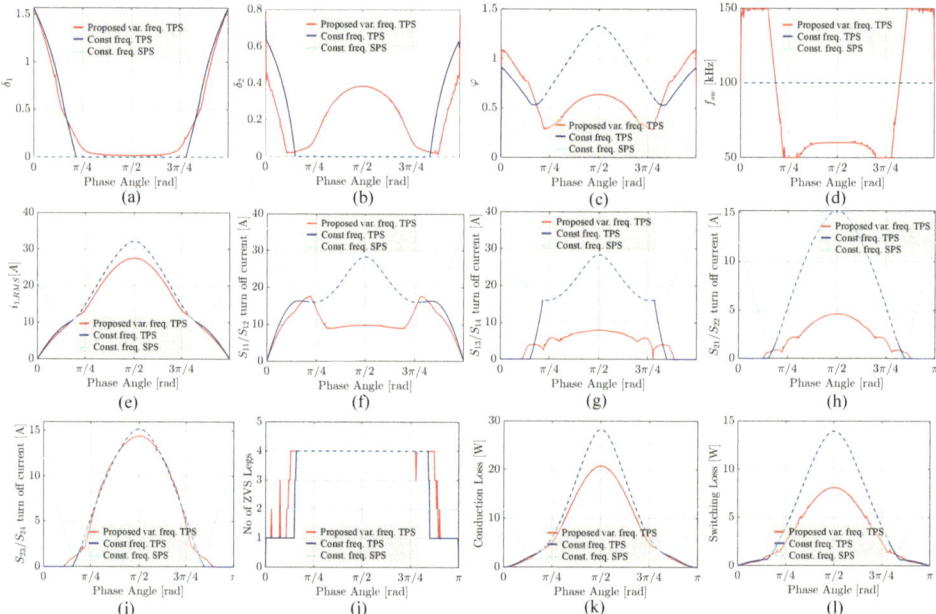

Fig. 3.13 DAB based ac-dc converter operation at 40 Vdc to 120 Vac power conversion mode at 500 W comparing variable frequency modulator and 100 kHz TPS modulator over half ac line period: **a** δ_1; **b** δ_2; **c** φ; **d** f_{sw}; **e** $i_{1,RMS}$; **f** S_{11}/S_{12} turn off current; **g** S_{13}/S_{14} turn off current; **h** S_{21}/S_{22} turn off current; **i** S_{23}/S_{24} turn off current; **j** Number of HF DAB legs undergoing ZVS; **k** Conduction loss; **l** Switching loss

(b) Another major benefit of the optimal modulator is the turn-off loss reduction of the HF DAB leg switches and extension of the ZVS range over an ac line cycle facilitating reduced switching losses in the switching network. It is observed from Fig. 3.12 that the number of HF DAB legs experiencing soft-switching increases over an ac line period based on the proposed control method compared to the constant frequency TPS control. This is fundamentally driven by the fact that for a given frequency, there is a certain dead band of operating region where ZVS is theoretically not possible [8]. It is possible to overcome that issue by adaptively shifting the location of that band by modulating the dc-ac converter switching frequency. Moreover, the average turn-off currents of the leading and lagging leg of the primary side HF H-bridge reduce by 31.2% and 63.8%, respectively, over an ac cycle under application of the proposed control method. Whereas the secondary side leading and lagging H-bridge legs experience 71.4% and 5.2% reduced average turn-off currents, respectively. The turn-off switching losses take a substantial portion ∼ 65% of the overall switching loss due to extended ZVS turn-on range. As a result, the proposed DAB converter achieves 7.5 W of switching loss reduction compared to the constant frequency modulated DAB, during the peak power transfer of 1000 W at $\Phi = 90°$.

To understand the comparison of the two methods with regard to EMI benefits, the frequency response characteristics of the output AC side currents are presented and compared in Fig. 3.14a–d. For linear loads, the frequency contents in the output currents would be the same as those in generated AC voltage. While Fig. 3.14a, b depicts the FFT results of the v_{ac} voltage attained from the variable frequency DAB control, the v_{ac} FFT results from TPS controlled DAB switching at 100 kHz are highlighted in Fig. 3.14c, d. Fig. 3.14a, c are generated from the FFT data received from the simulation of the dc-ac DAB for 40 Vdc to 120 Vac at 500 W power conversion mode; while Fig. 3.14b, d showcase the FFT data analytically derived from the proposed mathematical model. It is evident from Fig. 3.14a, b that the dominant HF voltage component in the v_{ac} appears at $2f_{sw}$ frequency, i.e., 200 kHz, when the DAB is switched at constant f_{sw} of 100 kHz. This appearance of the dominant HF component falls inside the EMI standard and will require substantial DM noise attenuation by the EMI filter stage. Furthermore, in this operating mode, the measured voltage THD from the simulation and the analytical result are found to be 2.31% and 2.18%, respectively. On the contrary, due to the variable f_{sw} operation of the optimal modulator, the FFT spectrum displays the HF voltage component of the AC side rectifier output spread over a wide range of frequencies ($2f_{sw} \in [100\,\text{kHz}, 300\,\text{kHz}]$). Factually, the maximum ac voltage ripple occurs during the peak power operation where f_{sw} is set at 62.5 kHz. Thus, the dominant HF voltage component in the FFT spectrum is noticed at $2f_{sw}$ or 125 kHz with the variable frequency modulator, which falls below the EMI standard start frequency of 150 kHz. The immediate next highest peak is located at 250 kHz (i.e., 4th harmonic of the switching frequency at the peak AC line) which shows a magnitude less than 0.1 V i.e., 100 dBuV, while the 200 kHz component at the

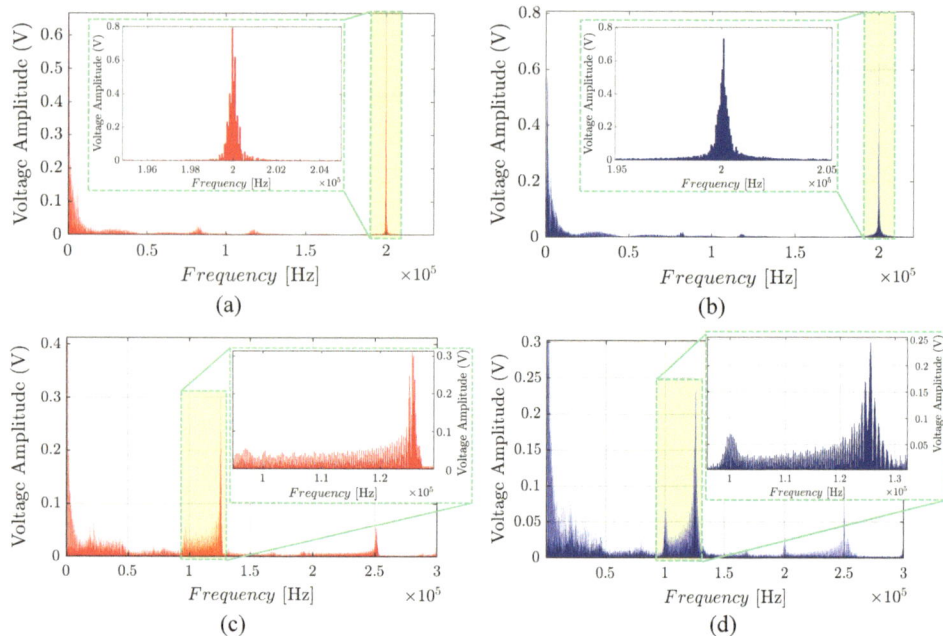

Fig. 3.14 FFT analysis of the dc-ac converter output voltage $v_{ac}(t)$ under (i) TPS DAB modulation with $f_{sw} = 100$ kHz [**a** simulation result; **b** analytical model] and (ii) variable frequency optimal TPS modulation [**c** simulation result; **d** analytical model]

fixed frequency modulator case is 0.8 V i.e., 118 dBuV. This result establishes the benefit of encountering lesser DM EMI noise peak amplitude employing the variable frequency modulator, fundamentally driven by the principle that the spectrum components are more distributed than being concentrated. With this method, the total THD of v_{ac} is observed to be 2.12% and 1.95% from the simulation and the analytical model, respectively.

3.6 Optimal Control Variable Synthesis Through Objective Function Optimization and Associated Control Implementation

3.6.1 Gradient Descent Approach

Notably, the optimal control variables need to be synthesized from the optimization routine framework with an update rate that is equal or faster than the controller main program execution frequency. Typically, the program execution frequency is kept the same as the switching frequency, which in turn imposes an upper bound on the execution time of the optimization framework. Gradient descent is a method for unconstrained mathematical optimization. It is a first-order iterative algorithm for finding a local minimum of a differentiable multivariate function. A model independent gradient descent search algorithm to be implemented in the TAB controller to achieve the accurate optimal operating point. The gradient descent algorithm is primarily chosen for this multivariable, single-objective optimization problem due to its computational efficiency that produces a stable error gradient and a stable convergence [9]. Also, the variable learning rate provides flexibility to the designer while choosing between the execution time and the optimization accuracy of the algorithm. The algorithm, as illustrated in Fig. 3.15, starts with defining the steady-state operating system parameters, voltage levels, and power demand. This is followed by the identification of the optimized PWM strategy according to the output load and voltage gain levels.

Thereafter, the values of the required duty cycle variables under the chosen PWM strategy are initialized as a vector $\delta(n)$ containing $[\delta_1(n), \delta_2(n), \delta_3(n)]$. At the same time, the primary control variables, i.e., the phase shifts $\phi_2(n)$ and $\phi_3(n)$, are adjusted by the implemented PI controller in order to meet the port voltage regulation and load demand.

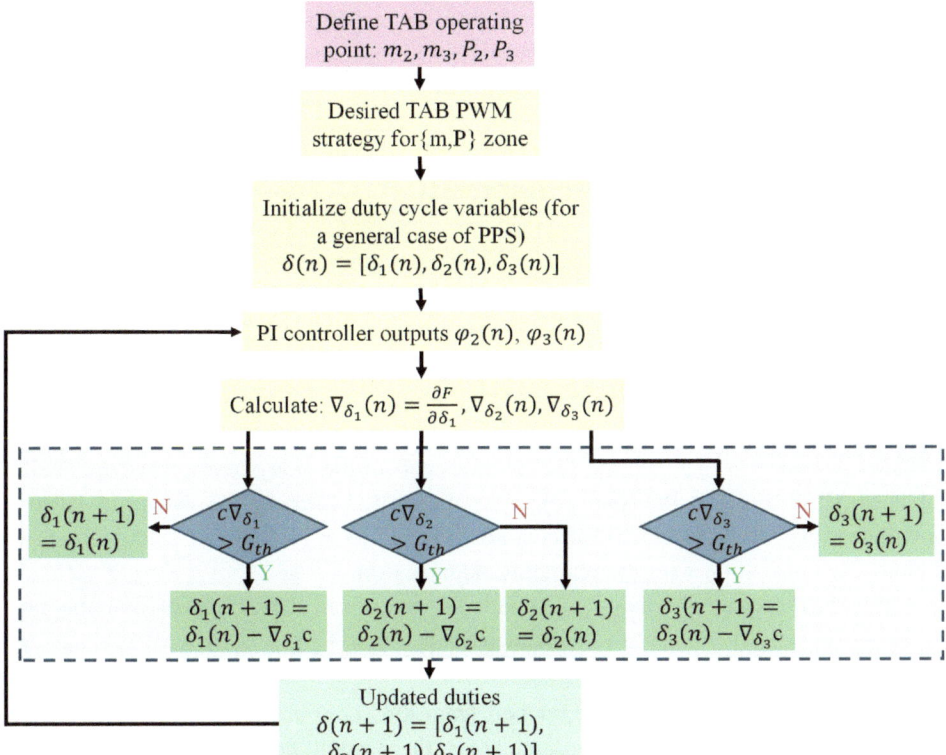

Fig. 3.15 Implemented algorithm to achieve the optimal control variables

Thus, the initial control variable set $[\delta(n), \phi(n)]$ and the objective function $[F(n)]$ containing total rms current or switching losses or net switching network losses are deduced for a particular sample time (n). Now, in the next step, the gradient of F, ∇F, is measured separately for change in individual duty cycle variables. These gradients are expressed as $\nabla\delta_1(n) = \frac{\partial F}{\partial\delta_1} = \frac{[F(n+1)-F(n)]}{\delta 1(n+1)-\delta 1(n)}$, $\nabla\delta_2(n)$, and $\nabla\delta_3(n)$. As long as these partial derivatives multiplied with the learning rate factor, c, are greater than the threshold G_{th}, the search algorithm continues updating the duty cycle variables one at a time. At every iteration of this algorithm step, the phase-shift variables are also settled by the PI controller. Finally, the optimum duty cycles are obtained whenever $c\nabla\delta_1, c\nabla\delta_2, c\nabla\delta_3$ become less than the threshold G_{th}. The values of c and G_{th} are chosen based on the tradeoff between target accuracy in reaching the optimum operating point operation and the sensitivity and execution time of the algorithm. As the learning rate increases, the optimum point can be reached faster at the cost of increased system dynamics at the time of change in load and required voltage gain. Considering the continuous, dynamic, and generic nature of this algorithm, the approach can be implemented in any multi-active bridge converter. The optimization can make use of MATLAB's *fmincon* function for nonlinear constrained optimization. To avoid convergence to local minima, a multi-start approach can be employed. This approach involves solving the optimization problem multiple times with different initial conditions, and the globally optimal solution is selected from the observed local minima values.

3.6.2 Look-Up Table Based Approach

The look up table would save the entries of the operating points and corresponding control variables. For a dc-dc or dc-ac DAB converter, the normalized voltage gain (m) and load power (P) are stored in a 2D M $\times$ N matrix, where M and N denote the number of trace points across the gain range and the load range, respectively. Based on accuracy requirements and control variable sensitivity against (m, P), the values of M and N are decided. For example, if a converter is operating in a gain range of [0.5, 1.5], a selection of M = 20 indicates that the gain resolution as part of the matrix entries is 0.05. A choice of N = 10 means that the load power resolution for forming the matrix is 10%. The possible values of each control variable are saved in M $\times$ N matrices, with element-wise one-to-one correlation with the operating point matrix.

Upon the detection of voltage gain and load power from sensor information, the indices (j, k) are calculated: $j = nint\left(1 + \frac{(m-m_{min})\times M}{(m_{max}-m_{min})}\right)$; $k = nint\left(1 + \frac{(P-P_{min})\times N}{(P_{max}-P_{min})}\right)$, where 'nint' is the nearest integer function.

The control variables are then fetched from the (j, k) entries of their corresponding matrices. For a variable frequency dc-ac DAB, there would be four control variables (i.e., frequency, phase, two bridge duty ratios) and hence four M $\times$ N control variable storage matrix. For (M = 20, N = 10) selection, the five matrices including the operating point matrix would consume a memory space for storing 1000 variables. For a floating point 32-bit definition of each variable, the memory space consumed for saving the matrix entries would be 4 kB. Now, if the load power resolution is set to 1% due to higher control variable sensitivity observed against load power, the value of N would be 100, requiring 40 kB memory space for storing the matrix elements. The scalability of LUT approaches in multiport converters is considered poor because they need to store an increasing amount of control variable data as the number of voltage gain and load power parameters grows. For example, in a 3-port TAB converter with two output ports and (M = 20, N = 10) for both ports, there would be 40,000 possible operating points and each of them would map to an optimal control variable set. For six control variables (i.e., frequency, two phase angles, three bridge duty ratios) into the picture, there would be six corresponding matrices, each with 40,000 entries, which combined with the operating point matrix would result in 280,000 variables to be stored in the memory. For 32-bit size per variable, the minimum memory requirement would be 1.12 MB. Notably, this exceeds the flash memory size of 1 MB supported by C2000 series TMS320F28379D microcontroller, one of the advanced dual-core digital signal processors (DSP). A very commonly used single-core C2000-series DSP is TMS320F28335, which supports only 512 kB flash memory and 68 kB RAM. Therefore, (M = 20, N = 10) selection for a TAB leveraging all control degrees of freedom is far from being executable in a F28335 DSP. In other words, for a given selection of MCU, the memory size directly determines the maximum order of switching modulation that is implementable for a particular M × N, or maximum divisions of (M, N) i.e., maximum realizable operating point resolution for a given modulation type. To reduce the memory allocation needs, the values of M and N need to be decreased with increased spacings between load and gain trace points, which however would result in inaccurate control variable generation and hence possible departure from optimum converter performance. Therefore, in a nutshell, although the LUT implementation is simple for controlling a 2-port converter, the realization would face space complexity as the number of output ports increases along with multivariable higher-order control requirement.

As an example study for controlling two-port dc-ac DAB, the fundamental control block diagram for the converter operation is presented in Fig. 3.16. It can be noticed that although the duty cycle and frequency variables are fetched from the look-up table, the phase-shift φ is dynamically modulated by a built-in PI controller to track the v_{ac} in a pure sine shape with 120 V RMS. This is done to circumvent any mismatch between the practical hardware implementation and analytically developed model of the DAB converter, based on which the optimal control datasets are calculated.

To store optimized control values δ_1, δ_2, and f_{sw} for ac output voltage phase angles (Φ) ranging from 1° to 90° in 2° interval corresponding to a specific power delivery $P_{ac, RMS}$, an LUT would require 3 rows and 45 columns, totaling 270 float variables. For 20 equidistant power levels, the total number of float variables would be 5400, requiring 21.6 kB memory. As the TMS320F28379D supports a RAM of 204 kB, it would be able

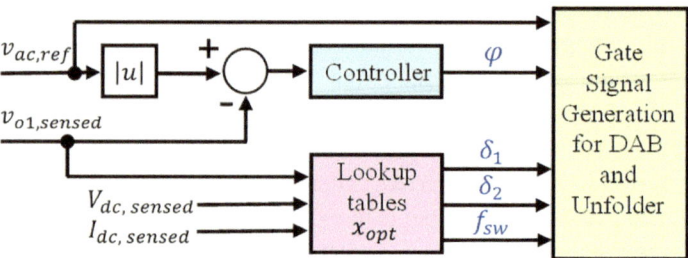

Fig. 3.16 Top level diagram showcasing proposed control implementation for the dc-ac DAB converter. [$V_{dc,sensed}$: DC voltage sensed on the input side, $I_{dc,sensed}$: DC current sensed on the input side, $v_{ac,ref}$: AC output voltage reference; $v_{o1,sensed}$: intermediate AC link voltage sensed]

to easily accommodate the memory requirements. However, for more extensive tables, with larger size and higher resolution in phase angle and power, using flash memory instead of RAM is recommended due to RAM's size limitations.

3.6.3   Multivariate Polynomial Regression Approach

As the LUT implementation becomes particularly challenging with exponentially worsening space complexity with an increasing dimension of modulation variables and also with the finely spaced array variable entries, its scalability to higher-order systems is heavily constrained. To overcome these issues, an alternate approach could be a hybrid-offline multivariate polynomial regression method that accounts for calculation of optimal variable sets for all possible voltage–power operating sets. Here, each control variable is expressed as a polynomial function of port gains and port power variables. The order of the polynomial is decided by studying the quantitative tradeoffs between regression accuracy and program execution time for computing the polynomial functions in DSP. If the execution time exceeds the converter switching cycle, the control loop functionality will be updated at a down-sampled rate of the sensed voltage and current that could lead to signal aliasing and loop instability issues. Hence, it becomes a necessity to limit the control execution time within one switching period. For example, the Nth order polynomials for constructing control variables (say, switching frequency and phase shift angle) are represented as follows, where **m** is the normalized voltage gain and **p** is the normalized port power (P/P_{rated}).

$$f_{sw} = \sum_{k=0}^{N}(\alpha_k m^k p^{N-k} + \overline{\alpha})$$

$$\varphi = \sum_{k=0}^{N} (\beta_k m^k p^{N-k} + \overline{\beta})$$

Polynomial coefficients (α_k, β_k) and intercepts $(\overline{\alpha}, \overline{\beta})$ are determined using least mean square error (LMSE) estimator based on offline optimization routine performed. The assumption behind the fact that the offline model would work well in real-time converter operation is that the system parameters (e.g., inductances, capacitances, device parameters) would remain unchanged during the operation phase. To further improve the optimization routine accuracy by accounting for system parameter variations under aging and environmental factors, artificial intelligence (AI)-driven techniques [10] can be employed and tied to the optimization framework.

3.7   Conclusions

This chapter presented a steady state modeling approach to a DAB DC-DC and DC-AC converter utilizing the frequency-domain GHA technique, which simplifies the multiple mode-based complex time-domain computation of transformer winding currents and power flow between the H-bridges. The comparative study of the available modulation techniques concludes the TPS modulation to be essential at light to medium load operation at non-unity gain to minimize the overall system loss and frequency modulation to be effective for a dc-ac or ac-dc DAB. This sets the context for the next chapters wherein we explore all possible combinations of frequency, phase, and duty modulation schemes in 3-port dc-dc-dc or dc-ac-dc isolated converters and their impact in loss minimization at a wide variety of operating points.

References

1. M. N. Kheraluwala, R. W. Gascoigne, D. M. Divan and E. D. Baumann, "Performance characterization of a high-power dual active bridge DC-to-DC converter," in *IEEE Transactions on Industry Applications*, vol. 28, no. 6, pp. 1294–1301, Nov.-Dec. 1992, https://doi.org/10.1109/28.175280.
2. A. K. Jain and R. Ayyanar, "Pwm control of dual active bridge: Comprehensive analysis and experimental verification," in *IEEE Transactions on Power Electronics*, vol. 26, no. 4, pp. 1215-1227, April 2011, https://doi.org/10.1109/TPEL.2010.2070519.
3. F. Krismer and J. W. Kolar, "Closed Form Solution for Minimum Conduction Loss Modulation of DAB Converters," in *IEEE Transactions on Power Electronics*, vol. 27, no. 1, pp. 174-188, Jan. 2012, https://doi.org/10.1109/TPEL.2011.2157976.
4. S. Wang, Z. Zheng, C. Li, K. Wang and Y. Li, "Time Domain Analysis of Reactive Components and Optimal Modulation for Isolated Dual Active Bridge DC/DC Converters," in *IEEE Transactions on Power Electronics*, vol. 34, no. 8, pp. 7143-7146, Aug. 2019, https://doi.org/10.1109/TPEL.2019.2897007.

5. G. G. Oggier, G. O. García and A. R. Oliva, "Switching Control Strategy to Minimize Dual Active Bridge Converter Losses," in *IEEE Transactions on Power Electronics*, vol. 24, no. 7, pp. 1826-1838, July 2009, https://doi.org/10.1109/TPEL.2009.2020902.
6. 650 V E-mode GaN Transistor GS-065–030–2-L, GaN Systems, Rev 220712 datasheet, 2009 [Revised 2021], URL: https://gansystems.com/wp-content/uploads/2022/07/GS-065-030-2-L-DS-Rev-220712.pdf
7. EPC2065 – Enhancement Mode Power Transistor, EPC, datasheet, 2021, URL: https://epc-co.com/epc/Portals/0/epc/documents/datasheets/EPC2065_datasheet.pdf
8. S. Dey and A. Mallik, "An Online-Optimized ZVS-Current Tracked Soft-Switching Modulation for Triple Active Bridge Converter," in IEEE Transactions on Power Electronics, https://doi.org/10.1109/TPEL.2024.3429278.
9. Pablo A. Parrilo, "Sum of Squares Optimization in the Analysis and Synthesis of Control Systems", lecture notes, http://www.mit.edu/~parrilo/pubs/talkfiles/Eckman.pdf.
10. S. Zhao, Y. Peng, Y. Zhang and H. Wang, "Parameter Estimation of Power Electronic Converters With Physics-Informed Machine Learning," in IEEE Transactions on Power Electronics, vol. 37, no. 10, pp. 11567-11578, Oct. 2022, https://doi.org/10.1109/TPEL.2022.3176468.

Switching Modulation Optimization in 3-Port TAB DC-DC and DC-AC Bidirectional Converters

4

4.1   Introduction to Three-Port Isolated Converters

In recent years, the proliferation of renewable energy resources and the growing demand for efficient energy management systems have underscored the need for compact, robust, and efficient multiport converters capable of seamlessly integrating diverse energy inputs and outputs [1]. Among these, triple-active bridge (TAB) converters have emerged as pivotal solutions due to their ability to facilitate bidirectional power flow among three ports while ensuring galvanic isolation and voltage matching capabilities [2–5]. TAB converters, a derivative of the dual-active bridge (DAB) family, are gaining substantial interest from both research and industry communities for their potential applications in electric vehicle chargers [6], energy routers, solid-state transformers [5], distributed renewable energy systems [7], and space power supplies [3]. Their architecture, which includes three actively controlled phase-shifted full H-bridges interconnected through a high-frequency three-winding transformer, enables efficient power transfer between ports with varying voltage and current requirements [5, 6]. This versatility is particularly advantageous in scenarios where multiple energy sources, storage devices, and loads need to be interconnected seamlessly.

Despite their advantages, TAB converters face several operational challenges that have spurred significant research efforts. One primary issue lies in their pulse-width modulation (PWM) strategies, especially under light load conditions and non-unity voltage gain scenarios. Traditional modulation techniques, such as dual phase shift (DPS) control, often result in increased conduction and switching losses due to higher transformer winding current RMS values, peaks, and loss of soft-switching [2–4, 8, 9]. These limitations necessitate the development of higher-order control strategies that can optimize power flow efficiency across all operating conditions.

© The Author(s), under exclusive license to Springer Nature Switzerland AG 2025
A. Mallik and S. Dey, *Switching Modulator Optimization in Isolated Power Converters*,
Synthesis Lectures on Power Electronics, https://doi.org/10.1007/978-3-031-81576-8_4

Current literature has explored various approaches to enhance the performance of TAB converters, investigating multi-variable modulation schemes that incorporate both phase shifts and duty cycle adjustments to minimize system losses [3, 4, 8, 9]. However, several challenges and drawbacks persist:

- **Optimal Duty Cycle Selection**: There is limited research on the optimal selection of the TAB bridge voltage duty cycles that ensures minimized loss operation for a wide range of load and gain conditions. While some studies, such as [2], introduced hybrid phase-duty parameters using fundamental harmonic approximation (FHA), FHA proves inefficient for estimating system behavior at lighter loads [4, 8] due to significant presence of higher order harmonic components in the TAB high-frequency inductor currents.
- **Control Variable Relationships**: Optimizing system performance requires careful synthesis of the relationships between control variables and current RMS values, peaks, and associated losses. FHA-based modeling often fails to capture these nuances, resulting in non-optimal control variables and less efficient system operation [3].
- **Zero-Voltage Switching (ZVS) Conditions**: Identification of ZVS conditions for all three full-bridge switching devices under hybrid phase-duty modulation, along with precise formulation of switching losses, remains underexplored. The loss of soft-switching, which significantly impacts efficiency at light loads and non-unity gain conditions, needs further investigation [4, 9, 10].
- **Optimized PWM Strategy Implementation**: Very few of the existing literature have explored the online hardware implementation methods of the loss optimized PWM control strategy for the TAB converter [3, 10–13]. Two primary methods are used: (a) offline determination of optimized control variables using FHA-based models stored in look-up tables (LUTs), which are not scalable due to large LUT sizes, and (b) gradient descent algorithms, which are complex and time-consuming. Furthermore, there is a lack of research on optimized closed-loop control strategies for DC-DC-AC TAB converters.

Addressing these challenges, this chapter aims to provide a comprehensive overview of switching modulation optimization strategies tailored specifically for 3-port TAB converters in both DC-DC and DC-AC configurations. The focus will be on understanding and resolving issues related to converter modeling, optimized modulation synthesis, and practical implementation. The chapter will present a novel approach using multivariate polynomial regression-based models for optimizing duty cycle control variables, designed to be easily implemented in digital signal processors (DSPs) for real-time control based on online load and AC voltage phase conditions. By exploring these themes, this chapter seeks to offer researchers, engineers, and practitioners a thorough understanding of the current state-of-the-art in TAB converter technology and to pave the way for more efficient and sustainable energy management solutions.

4.2   Modulation Optimization in a DC-DC TAB Converter

In a three-port TAB converter, the power flow among the ports as well as output voltage/current regulation are actively controlled by modulating the intra- and inter-bridge phase shifts of the gating signals and the switching frequency of the converter. The primary control parameters of a TAB converter are identified as: (a) phase-shifts of other bridge voltages with respect to primary (φ_2, φ_3), (b) duty cycles of the bridge voltages (δ_1, δ_2, δ_3), and (c) switching frequency (f_{sw}). These parameters are elucidated in Fig. 4.2, which illustrates the converter's bridge voltage waveforms under steady-state operation with phase-duty control. The intricate workings and mathematical modeling of a DC-DC TAB converter under steady-state conditions have been extensively covered in Chap. 2.

As delineated in Sect. 2.2.2 of Chap. 2, the desired power flow in a dual-output TAB converter, as defined by Eqs. (2.39) and (2.40), is achievable by utilizing two phase-shift control variables: φ_2 and φ_3, corresponding to the secondary and tertiary H-bridges. By solving these equations for the unknowns φ_2 and φ_3 while keeping the duty cycle variables (δ_1, δ_2, δ_3) at 0, a singular solution $\{\varphi_2, \varphi_3\}$ is obtained for a specific voltage gain and load demand $\{m_2, m_3, P_2, P_3\}$ of the converter. This control via phase shifts between the input and output full bridge cells facilitates the regulation of power transfer between the ports using a fundamental TAB PWM modulation technique known as Dual Phase Shift (DPS) modulation. However, this method confines the operation to a single feasible point, limiting opportunities for loss optimization within such a two-variable control framework.

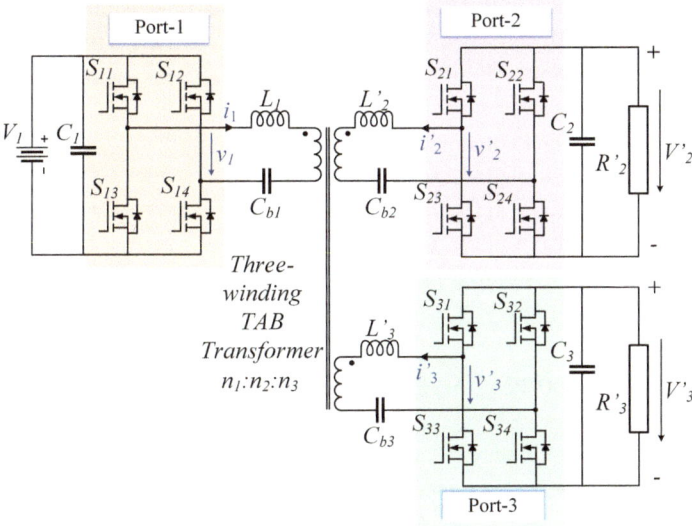

Fig. 4.1 TAB converter topology and phase shifts of the individual half-bridge gating signals

Fig. 4.2 Full-bridge output voltages in relation with the gate signal phase displacements of the individual half-bridges

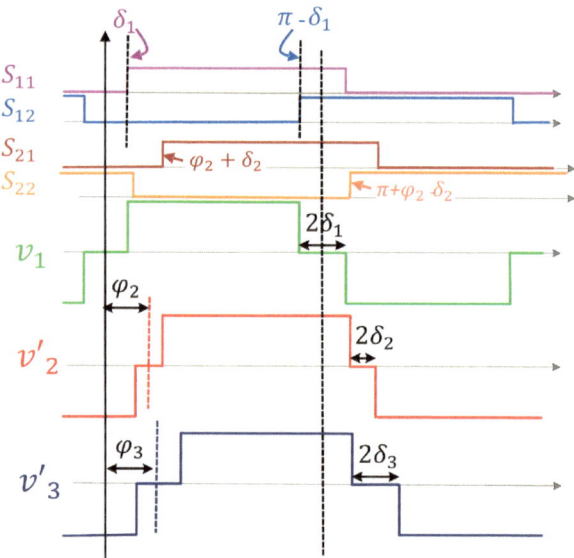

To introduce greater flexibility in controlling the power flow within a TAB converter, incorporating one or more duty cycles ($\delta_1, \delta_2, \delta_3$) of the bridge voltages as control variables is advantageous. Integrating additional duty cycle variables to satisfy specific load demands complicates the PWM modulation strategy but expands the set of potential control variables. This complexity enables the exploration of TAB loss optimization by selecting optimal control variables from the expanded set of feasible solutions. The PWM modulation techniques of a TAB converter, depending on the number of control degrees of freedom and their combinations, are broadly classified into eight distinct categories:

(i)   Dual Phase Shift (DPS): $\{\varphi_2, \varphi_3\}$
(ii) Triple Phase Shift—1 (TPS-1): $\{\delta_1, \varphi_2, \varphi_3\}$
(iii) Triple Phase Shift—2 (TPS-2): $\{\delta_2, \varphi_2, \varphi_3\}$
(iv) Triple Phase Shift—3 (TPS-3): $\{\delta_3, \varphi_2, \varphi_3\}$
(v)   Quad Phase Shift—1 (QPS-1): $\{\delta_2, \delta_3, \varphi_2, \varphi_3\}$
(vi) Quad Phase Shift—2 (QPS-2): $\{\delta_1, \delta_3, \varphi_2, \varphi_3\}$
(vii) Quad Phase Shift—3 (QPS-3): $\{\delta_1, \delta_2, \varphi_2, \varphi_3\}$
(viii) Penta Phase Shift (PPS): $\{\delta_1, \delta_2, \delta_3, \varphi_2, \varphi_3\}$.

Further, the switching frequency (f_{sw}) of the converter can also be actively adjusted alongside phase shift and duty ratios, generating a 6 variable control scheme, variable frequency PPS Modulation $\{\delta_1, \delta_2, \delta_3, \varphi_2, \varphi_3, f_{sw}\}$. Among these techniques, DPS is the simplest, whereas PPS requires the most computational effort due to its maximum degrees of freedom. Therefore, it's important to carefully select the appropriate technique that ensures efficient power conversion by minimizing the losses in the switching network across the wide range of output voltage gains and loads, while avoiding excessive computational demands on the digital implementation platform.

Now, in order to synthesize the optimal values of the TAB control variables for a specific operating gain and load condition that minimizes various system losses, the relations between the loss functions and the control and system parameters need to be precisely established. Moving forward, a multivariable multi constrained loss optimization problem needs to be formulated that results in the optimal control variable data set for given inputs of voltage gains and load conditions for the TAB ports.

4.2.1 Loss Function Formulation

The major losses in a TAB switching network comprising of the high-frequency switching semiconductor devices can be broadly classified into two categories such as conduction loss, and switching loss. The cost function for minimization in the multivariable optimization problem is oftentimes correlated with the different types of power losses or with the total loss in the system. The analytical formulations of the above-mentioned major loss components and their related cost functions are discussed below.

(i) Conduction Loss

The optimization of the triple active bridge (TAB) based three-port power converter efficiency is approached targeting the conduction loss minimization in few literatures [2, 3]. In these works, the objective function for optimization is formulated as the sum of winding RMS current ($i_{j,RMS}$) squares and can be shown as,

$$F_{RMS\_currents}(\delta_j, \varphi_j, f_{sw}) = \sum_{j=1}^{3} i_{j,RMS}^2 \tag{4.1}$$

where $i_{j,RMS}$ is normalized jth winding current RMS and can be formulated using the GHA model based synthesized expressions of the winding currents in a TAB converter, formulated in Chap. 2 and are also given in (4.2)–(4.4).

$$
i_{1,rms}^2 = \frac{8V_1^2}{\pi^2\omega^2} \cdot \sum_{k=1}^{2m+1} \frac{1}{k^4} \left[d_{k1}^2 \left(\frac{1}{L_{12}^2} + \frac{1}{L_{13}^2} + \frac{2}{L_{12}L_{13}} \right) + \frac{m_2^2 d_2^2}{L_{12}^2} + \frac{m_3^2 d_3^2}{L_{13}^2} \right.
$$
$$
\left. - 2d_{k1}d_{k2}m_2 \cos(k\varphi_2) \left(\frac{1}{L_{12}^2} + \frac{1}{L_{12}L_{13}} \right) \right.
$$

$$
- 2d_{k1}d_{k3}m_3 \cos(k\varphi_3)\left(\frac{1}{L_{13}^2} + \frac{1}{L_{12}L_{13}}\right)
$$

$$
+ 2m_2m_3d_{k2}d_{k3} \cos\{k(\varphi_2 - \varphi_3)\}\left(\frac{1}{L_{12}L_{13}}\right) \tag{4.2}
$$

$$
i_{2,rms}^2 = \frac{8V_1^2}{\pi^2\omega^2} \cdot \sum_{k=1}^{2m+1} \frac{1}{k^4}\left[(m_2d_{k2})^2\left(\frac{1}{L_{12}^2} + \frac{1}{L_{23}^2} + \frac{2}{L_{12}L_{23}}\right) + \frac{d_{k1}^2}{L_{12}^2} + \frac{(m_3d_{k3})^2}{L_{23}^2}\right.
$$

$$
- 2d_{k1}d_{k2}m_2 \cos(k\varphi_2)\left(\frac{1}{L_{12}^2} + \frac{1}{L_{12}L_{23}}\right)
$$

$$
- 2m_2m_3d_{k2}d_{k3} \cos\{k(\varphi_2 - \varphi_3)\}\left(\frac{1}{L_{23}^2} + \frac{1}{L_{12}L_{23}}\right)
$$

$$
\left. + 2d_{k1}d_{k3}m_3 \cos(k\varphi_3)\left(\frac{1}{L_{12}L_{23}}\right)\right] \tag{4.3}
$$

$$
i_{3,rms}^2 = \frac{8V_1^2}{\pi^2\omega^2} \cdot \sum_{k=1}^{2m+1} \frac{1}{k^4}\left[(m_3d_{k3})^2\left(\frac{1}{L_{13}^2} + \frac{1}{L_{23}^2} + \frac{2}{L_{13}L_{23}}\right) + \frac{d_1^2}{L_{13}^2} + \frac{(m_2d_{k2})^2}{L_{23}^2}\right.
$$

$$
- 2d_{k1}d_{k3}m_3 \cos(k\varphi_3)\left(\frac{1}{L_{13}^2} + \frac{1}{L_{13}L_{23}}\right)
$$

$$
- 2m_2m_3d_{k2}d_{k3} \cos\{k(\varphi_3 - \varphi_2)\}\left(\frac{1}{L_{23}^2} + \frac{1}{L_{13}L_{23}}\right)
$$

$$
\left. + 2d_{k1}d_{k2}m_2 \cos(k\varphi_2)\left(\frac{1}{L_{13}L_{23}}\right)\right] \tag{4.4}
$$

To verify the effectiveness of introducing duty cycle control variables for minimizing the cost function of (4.1), the gradient of the objective function $F = F_{RMS\_currnets}(\delta_j, \varphi_j)$ with respect to the duty cycle δ_1 is derived, as shown in Eq. (4.5).

$$
\nabla_{\delta_1}F = \frac{\partial F}{\partial \delta_1} = \frac{\partial i_{1,RMS}^2}{\partial \delta_1} + \frac{\partial i_{2,RMS}^2}{\partial \delta_1} + \frac{\partial i_{2,RMS}^2}{\partial \delta_1}
$$

$$
= \frac{48V_1^2}{\pi^2\omega^2L^2} \cdot \sum_{k=1}^{2m+1} \frac{\sin(k\delta_1)}{k^3} \cdot [m_2\cos(k\delta_2)\cos(k\varphi_2)
$$

$$
+ m_3\cos(k\delta_3)\cos(k\varphi_3) - 2\cos(k\delta_1)] \tag{4.5}
$$

Under the application of TPS-1 modulation, where $\delta_2 = \delta_3 = 0$, the gradient expression simplifies to Eq. (4.6).

$$\nabla_{\delta_1} F\big|_{\delta_2=\delta_3=0} = \frac{48V_1^2}{\pi^2\omega^2 L^2} \cdot \sum_{k=1}^{2m+1} \frac{\sin(k\delta_1)}{k^3}[m_2\cos(k\varphi_2) + m_3\cos(k\varphi_3) - 2\cos(k\delta_1)]$$

$$(4.6)$$

For simplicity of analysis, considering only fundamental harmonic modeling, i.e., m = 1, the gradient ∇F_{δ_1} under TPS-1 operation is restructured as follows:

$$\nabla_{\delta_1} F\big|_{\delta_2=\delta_3=0} = \frac{48V_1^2}{\pi^2\omega^2 L^2} \cdot \sin(\delta_1)[m_2\cos(\varphi_2) + m_3\cos(\varphi_3) - 2\cos(\delta_1)] \qquad (4.7)$$

Evaluating the gradient at $\delta_1 = 0^+$, we obtain the following:

$2\cos(\delta_1) \approx 2 > m_2\cos(\varphi_2) + m_3\cos(\varphi_3)$ for any φ_2 and φ_3 when $m_2, m_2 < 1$. Hence, for $m_2, m_2 \in [0, 1)$, a negative gradient, i.e., $\nabla_{\delta_1} F\big|_{\delta_1=0^+;\delta_2=\delta_3=0} < 0$ holds universally true. Therefore, it can be deduced that employing the TPS-1 modulation technique, as opposed to the basic DPS-based approach, contributes to minimizing the objective function within the specified operating parameters (m_2, m_3, P_2, P_3). This observation also underscores the inadequacies of DPS-based modulation, which does not achieve the lowest sum of mean square currents using solely phase shift-based control across a broad {m, P} operational range.

Additionally, to substantiate the need for more advanced modulation methods like QPS-3 compared to TPS-1, we analyze the gradient of ∇F_{δ_1} with respect to δ_2, as detailed in Eq. (4.8).

$$\nabla_{\delta_2}(\nabla_{\delta_1} F) = \frac{48V_1^2}{\pi^2\omega^2 L^2} \cdot \sum_{k=1}^{2m+1} \frac{\sin(k\delta_1)}{k^2}[-m_2\sin(k\delta_2)\cos(k\varphi_2)] \qquad (4.8)$$

Applying m = 1 in Eq. (4.8), we find:

$$\nabla_{\delta_2}(\nabla_{\delta_1} F) = -\frac{48V_1^2}{\pi^2\omega^2 L^2} \cdot m_2\sin(\delta_1)\sin(\delta_2)\cos(\varphi_2) \qquad (4.9)$$

For $\delta_2 > 0^+$, $\nabla_{\delta_2}(\nabla_{\delta_1} F) < 0$ holds true at a specific $\delta_1(> 0)$. Thus, at a given δ_1, QPS-3 modulation will more effectively minimize F compared to TPS-1. This indicates that using higher order PWM modulation strategies can be beneficial for reducing conduction loss, although this benefit comes at the expense of increased computational complexity.

However, the objective function given in (4.1) does not fully capture the conduction losses in the components of the TAB power stage because these losses also depend on the on-state channel resistance ($R_{DS,on}$) of the devices, which varies depending on the devices selected. Therefore, a more precise conduction loss equivalent objective function can be formulated as follows.

$$F_{cond\_loss}(\delta_j, \varphi_j, f_{sw}) = \sum_{j=1}^{3} i_{j,RMS}^2 R_j \qquad (4.10)$$

where, R_j = normalized $R_{DS,on}$ of the jth port switches. But (4.1) offers a more generalized objective function that remains effective regardless of the device types used. It is particularly useful in situations where the secondary on-state resistances, after reflection in the primary side, become nearly equal to the primary on-state resistances.

(ii) **Switching Loss**

The switching loss in a TAB converter depends on the peak winding currents at each switching instant, its corresponding bridge's dc link voltage, the ZVS conditions and the semiconductor device parasitics. The winding current values at the switching instants can be derived from the instantaneous winding current expressions of the TAB converter, given as (4.11).

$$i_i(t) = \sum_{k=1,3,5,\dots}^{\infty} \left(A_{i,k} \cos(k\omega_s t) + B_{i,k} \sin(k\omega_s t) \right). \tag{4.11}$$

where the harmonic order dependent coefficients $A_{i,k}$ and $B_{i,k}$ are:

$$A_{i,k} = \frac{4}{\pi \omega k^2} \left[\sum_{\substack{j=1 \\ j \neq i}}^{3} \frac{1}{L_{ij}} \{ -V_i d_{ki} \cos(k\varphi_i) + V_j d_{kj} \cos(k\varphi_j) \} \right] \quad \text{and} \quad B_{i,k} =$$

$\frac{4}{\pi \omega k^2} \left[\sum_{\substack{j=1 \\ j \neq i}}^{3} \frac{1}{L_{ij}} \{ -V_i d_{ki} \sin(k\varphi_i) + V_j d_{kj} \sin(k\varphi_j) \} \right]$. Due to the half-cycle symmetric current waveshapes as observed from (4.11), the turn-on current of any switch will be exactly same as the turn-off current but opposite in direction. Keeping this relation in mind the switching loss equivalent cost functions for a TAB are formulated while imposing the ZVS conditions, derived earlier in Chap. 2.

Due to the major dependency of switching loss on the switch currents at switching instants, a simple cost function that relates to the switching loss in a TAB is formulated in (4.12), and is expressed as the sum of the peak currents during the hard switching (turn-off and turn-on) transitions.

$$F_{sw\,pk}\left(\delta_j, \varphi_j, f_{sw}\right) = \begin{cases} \sum_{j=1}^{3} \left[4|i_j(\varphi_j)| + 4|i_j(\varphi_j)| \cdot (hard(j)) \right]; \, \delta_j = 0 \\ \sum_{j=1}^{3} \sum_{k=1}^{2} \left[2|i_{j,k}(\tau_{j,k})| + 2|i_{j,k}(\tau_{j,k})| \cdot (hard(j,k)) \right]; \, \delta_j > 0 \end{cases} \tag{4.12}$$

where $hard(j, k)$ can be 1 or 0, depending on if the kth (k = 1, 2) half-bridge of the jth port (j = 1, 2, 3) is experiencing hard turn-on, or not. In the context of FET turn-on, the condition hard(j) = 0 occurs for a complete soft-switching or for a valley switching case, wherein the rising edge of gate-source voltage appears right at the valley of the falling drain-source voltage waveform. In a perfect valley switching case, V-I product-associated loss is zero, while the major portion of the loss is attributed to CV^2 losses based on the residue voltage across the FET body capacitor, as detailed in (4.17). For all other cases including pure hard switching due to unfavorable current polarity or partial non-valley soft switching due to insufficient tank current or inadequate deadtime, hard(j) $= 1$ holds true. The turn-off current of kth ($k = 1, 2$) leg half-bridge of the jth port is shown as $|i_{j,k}(\tau_{j,k})|$, where $\tau_{j,k}$ denotes the switching instants as $\omega t = \varphi_j + \delta_j$ for leading leg, $\omega t = \pi + \varphi_j - \delta_j$ for lagging leg and can be calculated using (4.11). For ease of understanding, the synthesis of the switching loss equivalent function $F_{swpk}(\delta_j, \varphi_j, f_{sw})$ and ZVS function $ZVS(j, k)$ is presented in Fig. 4.3 in a form of flowchart. Here, $ZVS(j, k)$ can be 1 or 0, depending on if the kth ($k = 1, 2$) half-bridge of the jth port (j = 1, 2, 3) is undergoing complete soft turn-on, or not. The ZVS conditions used in the algorithm are determined from Table 4.1 as described in Chap. 2. Although $F_{swpk}(\delta_j, \varphi_j)$ does not model the actual switching loss in the converter, it is quantified in such a way that minimization of the function tries to minimize the overall peak switching currents while avoiding hard-switching events.

Further, a better modeling of the switching loss function is also possible that tries to model the loss incurred by the overlap between current through the device and voltage across the device during any switching event. Thus, such a model should consider the port voltage levels and switching device parameters such as gate-source capacitance (C_{gs}) and body capacitance (C_{oss}) into account alongside the switching peak currents and the hard-switching conditions. Such a precise switching loss-oriented cost function is presented as,

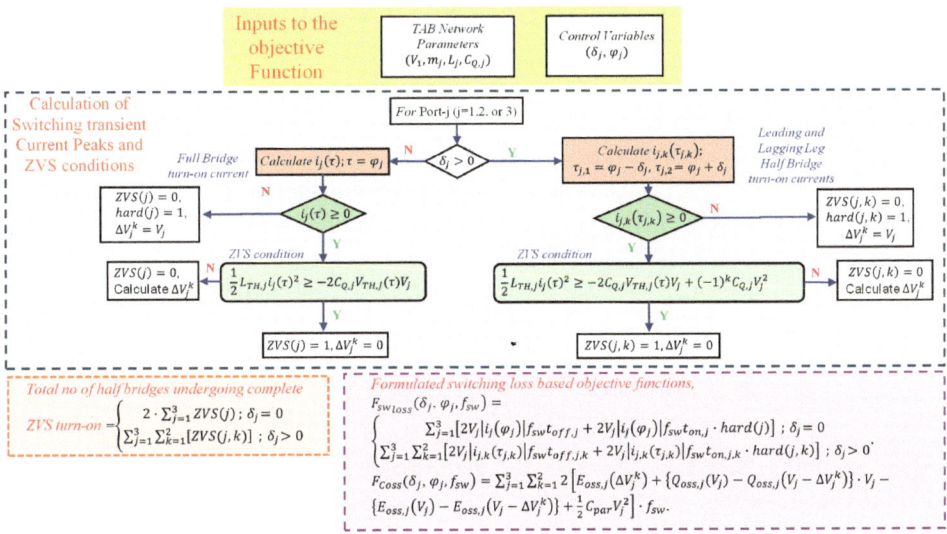

Fig. 4.3 Flowchart showing formulation of objective functions for switching loss optimization in a TAB network

Table 4.1 Summarized hard-switching and conditions for complete ZVS in TAB

Case	Switch undergoing turn-on event	Condition for hard-switching	Condition for complete ZVS	Switching time (τ)
1(a)	S_{j2} and S_{j3}	$i_j(\tau) < 0$	$\frac{1}{2}L_{TH,j}i_j(\tau)^2 \geq 2C_{Q,j}(V_j)v_{TH,j}(\tau)V_j$	$\varphi_j + \pi$
1(b)	S_{j1} and S_{j4}	$i_j(\tau) > 0$	$\frac{1}{2}L_{TH,j}i_j(\tau)^2 \geq -2C_{Q,j}(V_j)v_{TH,j}(\tau)V_j$	φ_j
2(a)	S_{j3}	$i_j(\tau) > 0$	$\frac{1}{2}L_{TH,j}i_j(\tau)^2 \geq$ $2C_{Q,j}(V_j)v_{TH,j}(\tau)V_j - C_{Q,j}(V_j)V_j^2$	$\varphi_j + \pi - \delta_j$
2(b)	S_{j1}	$i_j(\tau) > 0$	$\frac{1}{2}L_{TH,j}i_j(\tau)^2 \geq$ $-2C_{Q,j}(V_j)v_{TH,j}(\tau)V_j - C_{Q,j}(V_j)V_j^2$	$\varphi_j + \delta_j$
2(c)	S_{j2}	$i_j(\tau) > 0$	$\frac{1}{2}L_{TH,j}i_j(\tau)^2 \geq$ $2C_{Q,j}(V_j)v_{TH,j}(\tau)V_j + C_{Q,j}(V_j)V_j^2$	$\varphi_j + \pi - \delta_j$
2(d)	S_{j4}	$i_j(\tau) > 0$	$\frac{1}{2}L_{TH,j}i_j(\tau)^2 \geq$ $-2C_{Q,j}(V_j)v_{TH,j}(\tau)V_j + C_{Q,j}(V_j)V_j^2$	$\varphi_j - \delta_j$

$$F_{sw\,loss}(\delta_j, \varphi_j, f_{sw}) =$$
$$\begin{cases} \sum_{j=1}^{3}\left[2V_j\left|i_j(\varphi_j)\right|f_{sw}t_{off,j} + 2V_j\left|i_j(\varphi_j)\right|f_{sw}t_{on,j} \cdot hard(j)\right]; \delta_j = 0 \\ \sum_{j=1}^{3}\sum_{k=1}^{2}\left[2V_j\left|i_{j,k}(\tau_{j,k})\right|f_{sw}t_{off,j,k} + 2V_j\left|i_{j,k}(\tau_{j,k})\right|f_{sw}t_{on,j,k} \cdot hard(j,k)\right]; \delta_j > 0 \end{cases} \quad (4.13)$$

where f_{sw}, $t_{on,j,k}$ and $t_{off,j,k}$ denote the converter switching frequency, device turn-on and turn-off time of kth leg of jth TAB port, respectively. Here, $t_{on,j,k}$ and $t_{off,j,k}$ depend on the device parameters (C_{oss}, C_{iss}, C_{rss}) obtained from the datasheet and on the gate resistances ($R_{G,on}$ and $R_{G,off}$) and are calculated by following the relations given in (4.14) and (4.15), respectively [14].

$$t_{on,j,k} = R_{g,on,j}C_{iss,j}\ln\left(\frac{V_{dr,j} - V_{th,j}}{V_{dr,j} - V_{pl,j}}\right) + \frac{C_{rss,j}\left(V_j - R_{on,j}\left|i_{j,x}(\tau_{j,k,on})\right|\right)R_{g,on,j}}{V_{dr,j} - V_{pl,j}}; \quad (4.14)$$

$$t_{off,j,k} = R_{g,on,j}C_{iss,j}\ln\left(\frac{V_{dr,j} - V_{th,j}}{V_{dr,j} - V_{pl,j}}\right) + \frac{C_{rss,j}\left(V_j - R_{on,j}\left|i_{j,x}(\tau_{j,x,on})\right|\right)R_{g,on,j}}{V_{dr,j} - V_{pl,j}}. \quad (4.15)$$

Here, $V_{dr,j}$, $V_{th,j}$ and $V_{pl,j}$ represent the gate drive voltage, gate threshold voltage and plateau voltage of the switching MOSFET connected at port-j of the TAB converter.

Furthermore, the total number of switching legs undergoing soft turn-on in a TAB converter under a steady state operating point can be determined as (4.16).

$$F_{ZVS}(\delta_j, \varphi_j, f_{sw}) = \sum_{j=1}^{3} \sum_{k=1}^{2} ZVS(j, k) \tag{4.16}$$

$F_{ZVS}(\delta_j, \varphi_j, f_{sw})$ is also a possible cost function candidate, maximizing which ensures most number of HF switching legs in a TAB power stage will experience soft-switching for a specific load and gain condition.

Apart from the losses incurred by overlap between switching voltage and current during any switching event, the incomplete charge and discharge of the MOSFET's body capacitors (C_{oss}) also add on to the switching loss under and partial or full hard-switching event. From the discussion of energy dissipation during incomplete ZVS and hard switching events, delineated in Chap. 2, the C_{oss} related loss in a TAB converter can be estimated as,

$$F_{Coss}(\delta_j, \varphi_j, f_{sw}) = \sum_{j=1}^{3} \sum_{k=1}^{2} 2\Big[E_{oss,j}\Big(\Delta V_j^k\Big) + \Big\{ Q_{oss,j}(V_j) - Q_{oss,j}\Big(V_j - \Delta V_j^k\Big) \Big\} \Delta V_j$$

$$- \Big\{ E_{oss,j}(V_j) - E_{oss,j}\Big(V_j - \Delta V_j^k\Big) \Big\} + \frac{1}{2} C_{par} V_{par}^2 \Big] \cdot f_{sw}. \tag{4.17}$$

Here, $E_{oss,j}\Big(\Delta V_j^k\Big)$ denotes the energy stored in the nonlinear capacitance C_{oss} of the MOSFET connected at k-th switching leg of j-th bridge due to the presence of ΔV_j^k voltage across it during its turn-on instant. This residue drain-source voltage ΔV_j^k at turn-on instant can be found out by solving the energy balance as shown in Chap. 2 (Ref. (2.72)). For any drain-source voltage V_{DS}, the stored energy at C_{oss} is $E_{oss}(V_{DS}) = \int_0^{V_{DS}} C_{oss}(v) \cdot v dv = \frac{1}{2} C_E(V_{DS}) \cdot V_{DS}^2$ and can be calculated from the $C_{oss} - V_{DS}$ plot given in the device datasheet. Similarly, $Q_{oss}(V_{DS})$ denotes the amount of charge stored in C_{oss} due to V_{DS} voltage present across it (see Fig. 4.4) and can be calculated as, $Q_{oss}(V_{DS}) = \int_0^{V_{DS}} C_{oss}(v) \cdot dv = C_Q(V_{DS}) \cdot V_{DS}$. The C_{par} is the parasitic switching node capacitance that appears due to the printed circuit board design, parasitic capacitances of the transformer etc., and needs to be charged and discharged everytime during any switching commutation. V_{par} is the voltage across C_{par}, which for a transformer intra-winding capacitance will be $2V_j$ for an H-bridge with no duty modulation and be V_j for a duty-modulated H-bridge. In case of complete soft-switching transition, ΔV_j^k will be 0, while, for a complete hard-switching, $\Delta V_j^k \to V_j$.

Finally, a cost function $F_{sw\_loss\_total}(\delta_j, \varphi_j, f_{sw})$ that accurately models the loss in a TAB switching network due to the switching action can be derived as a sum of $F_{swloss}(\delta_j, \varphi_j, f_{sw})$ and $F_{Coss}(\delta_j, \varphi_j, f_{sw})$.

$$F_{sw\_loss\_total}(\delta_j, \varphi_j, f_{sw}) = F_{swloss}(\delta_j, \varphi_j, f_{sw}) + F_{Coss}(\delta_j, \varphi_j, f_{sw}) \tag{4.18}$$

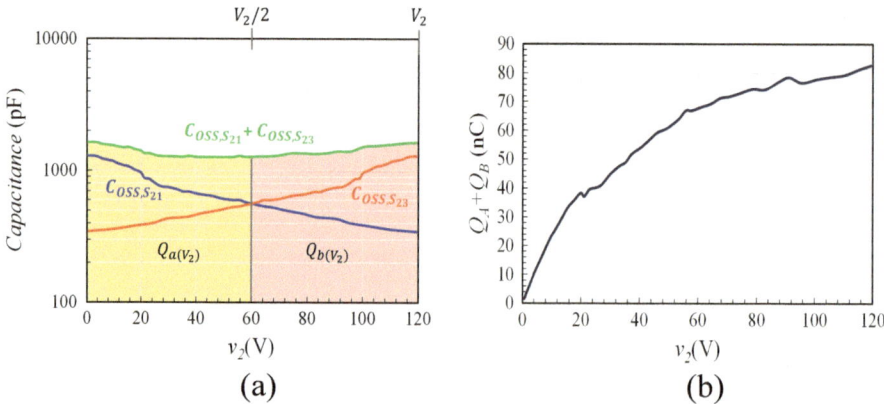

Fig. 4.4 Variation of **a** C_{oss} and **b** its stored charge with varying v_{DS} for EPC 2215 GaN MOSFET

Such objective function can be used in order to optimize and model the converter switching loss more accurately compared to (4.12); however, the optimal control variable set will become more dependent on the converter power stage parameters.

Thus, different objective functions that corelates to different loss components within the TAB switching network are derived and are presented in Table 4.2.

4.2.2 Control Optimization Problem Formulation

Upon formulating the different loss equivalent objective functions for loss optimization in a TAB switching network, it is time to define the optimization problem that will generate the optimal control variables for any steady state converter operating point. From the perspective of the TAB modulation optimization, we need to first establish the system constraints, decision variables and objective function for the problem definition.

- Decision Variables

The modulation of the TAB converter can be generalized as the highest order control system employing a variable frequency PPS modulation strategy. All the other lower order modulation techniques can be easily represented as a subset of such generalized modulation. Thus, the decision variable set can be defined as $x = \{\delta_1, \delta_2, \delta_3, \varphi_2, \varphi_3, f_{sw}\}$.

- Constraints

The specific load demands at the output ports of the TAB converter impose the primary non-linear constraints on the optimization problem. If the port-2 and port-3 are treated as

Table 4.2 Summarized formulated objective functions for various loss optimizations in TAB

S. No.	Target loss type	Formulated objective function for minimization
1(a)	*Switching loss*	$F_{sw\,pk}(\delta_j,\varphi_j,f_{sw})$ $= \begin{cases} \sum_{j=1}^{3}[4|i_j(\varphi_j)| + 4|i_j(\varphi_j)| \cdot hard(i)]; \delta_j = 0 \\ \sum_{j=1}^{3}\sum_{k=1}^{2}[2|i_{j,k}(\tau_{j,k})| + 2|i_{j,k}(\tau_{j,k})| \cdot hard(j,k)]; \delta_j > 0 \end{cases}$
1(b)		$F_{sw\,loss}(\delta_j,\varphi_j,f_{sw})$ $= \begin{cases} \sum_{j=1}^{3}[2V_j|i_j(\varphi_j)|f_{sw}t_{off\,j} + 2V_j|i_j(\varphi_j)|f_{sw}t_{on,j} \cdot hard(j)]; \delta_j = 0 \\ \sum_{j=1}^{3}\sum_{k=1}^{2}[2V_j|i_{j,k}(\tau_{j,k})|f_{sw}t_{off\,j,k} + 2V_j|i_{j,k}(\tau_{j,k})|f_{sw}t_{on,j,k} \cdot hard(j,k)]; \delta_j > 0 \end{cases}$
1(c)		$F_{Coss}(\delta_j,\varphi_j,f_{sw}) =$ $\sum_{j=1}^{3}\sum_{k=1}^{2} 2\left[E_{oss,j}\left(\Delta v_j^k\right) + \left\{Q_{oss,j}(V_j) - Q_{oss,j}\left(V_j - \Delta v_j^k\right)\right\}\Delta V_j - \left\{E_{oss,j}(V_j) - E_{oss,j}\left(V_j - \Delta V_j^k\right)\right\} + \frac{1}{2}C_{par}V_j^2\right] \cdot f_{sw}$
2(a)	*Conduction loss*	$F_{RMS\_currents}(\delta_j,\varphi_j,f_{sw}) = \sum_{j=1}^{3} i_{j,RMS}^2; i_{j,RMS} =$ normalized winding current RMS
2(b)		$F_{cond\_loss}(\delta_j,\varphi_j,f_{sw}) = \sum_{j=1}^{3} i_{j,RMS}^2 R_j; R_j =$ normalized $R_{DS(ON)}$
3(a)	*Total converter loss*	$F_{total\_loss}(\delta_j,\varphi_j,f_{sw}) = F_{sw\_loss} + F_{Coss} + F_{cond\_loss} + F_{TF\_loss} + F_{cap\_loss}; F_{TF\_loss} =$ Losses at Transformer and $F_{cap\_loss} =$ Losses at DC link Cap
3(b)	*Total converter loss while maximizing ZVS*	(i) $F_{total\_loss}(\delta_j,\varphi_j,f_{sw})$ (ii) $TotalZVSCount$

the output ports and the total load connected to them are P_2, and P_3, respectively, then the power flow constraints can be formed by equating each of them with the GHA model derived port power expressions, as given in (4.19) and (4.20).

$$P_2 = \frac{8P_{base}}{\pi^2} \sum_{k=1}^{2m+1} \frac{1}{k^3} \left[\frac{3m_2L_1}{L_{12}} d_{k1}d_{k2} \sin(k\varphi_2) + \frac{3m_2m_3L_1}{L_{23}} d_{k2}d_{k3} \sin\{k(\varphi_2 - \varphi_3)\} \right]$$

(4.19)

$$P_3 = \frac{8P_{base}}{\pi^2} \sum_{k=1}^{2m+1} \frac{1}{k^3} \left[\frac{3m_3L_1}{L_{13}} d_{k1}d_{k3} \sin(k\varphi_3) + \frac{3m_2m_3L_1}{L_{23}} d_{k2}d_{k3} \sin\{k(\varphi_3 - \varphi_2)\} \right]$$

(4.20)

The power flow equations are already formulated in Chap. 1, where $P_{base} = \frac{V_1^2}{2\pi f_{sw}L_1}$, $d_{k1} = \cos(k\delta_1)$ and $d_{k2} = \cos(k\delta_2)$.

Other than the non-linear equality constraints, the upper and lower bounds of the decision variables will also impose linear constraints on the optimization:

$$0 \leq \delta_k \leq \frac{\pi}{2};$$

(4.21)

$$-\frac{\pi}{2} \leq \varphi_k \leq \frac{\pi}{2};$$

(4.22)

$$f_{sw,min} \leq f_{sw} \leq f_{sw,max}.$$

(4.23)

The upper and lower bounds of these variables can be chosen by the designer. For example, if a constant frequency (e.g. 100 kHz) PPS modulation optimization is aimed, then f_{sw} should be made constant at 100 kHz $= f_{sw,min} = f_{sw,max}$.

- Cost Function

According to the user need, the objective function for the optimization algorithm can be defined in various ways targeting different loss elements in the HF TAB stage such as conduction, switching loss or ZVS extension. Thus, the cost function $f_{cost}(x)$ can be defined as

$$f_{cost}(x) = F_{sw\,loss}\left(\delta_j, \varphi_j, f_{sw}\right) \text{ or } F_{cond_{loss}}\left(\delta_j, \varphi_j, f_{sw}\right) \text{ or } F_{sw\_loss\_total}\left(\delta_j, \varphi_j, f_{sw}\right)$$
$$\text{or } F_{sw\,pk}\left(\delta_j, \varphi_j, f_{sw}\right) \text{ or } F_{RMS\_currents}\left(\delta_j, \varphi_j, f_{sw}\right) \text{ or } -F_{ZVS}\left(\delta_j, \varphi_j, f_{sw}\right).$$

(4.24)

Thus, for a specific converter operating condition, defined by the output ports' voltage gain and load conditions (m_2, m_3, P_2, P_3), the multi-variable, multi-constrained, and single objective optimization problem can be solved to synthesize the optimized control variable set $x_{opt} = \{\delta_1, \delta_2, \delta_3, \varphi_2, \varphi_3, f_{sw}\}$ that leads to the least cost function value for a particular (m_2, m_3, P_2, P_3) set. This optimization can be easily performed using any numerical non-linear optimization methods employing software tools such as MATLAB, python and Mathematica.

4.2.3 Case Study: PWM Strategy Optimization for a DC-DC TAB

In this case study, we explore the optimization of a PWM strategy for a Three-Port (TAB) DC-DC converter. This converter has a three-port design with an input voltage of 160 V and adjustable output voltages ranging from 100–140 V and 16–28 V, with each output port capable of handling 400 W. These voltage levels are specifically designed for space station power supply systems. The detailed circuit parameters of the TAB converter are summarized in Table 4.3. Using this example, we illustrate the control optimization approach.

During formulating TAB modulation optimization problem in the last section, we discovered that employing a six-variable control approach, specifically the variable frequency Phase-Shifted Pulse Width Modulation (PPS) technique, consistently yields optimal conduction loss in a TAB control architecture. However, implementing a variable frequency PPS control across the entire range of output voltage gain and load conditions presents challenges for the TAB controller due to the computational complexity and large-memory requirements of look-up tables. Consequently, this study focuses on identifying an optimized and favorable constant frequency (100 kHz) PWM control strategy tailored to the port voltage gain and power levels (m_2, m_3, P_2, P_3). The cost function for optimization

Table 4.3 Example TAB circuit parameters

Circuit parameters	Values
Input DC voltage (V_1)	160 V
Secondary output port voltage (V_2)	100–140, 120 V (nominal)
Secondary output port rated power (P_2)	400 W
Tertiary output port voltage (V_3)	16–28, 22 V (nominal)
Tertiary output port rated power (P_3)	400 W
Transformer turns ratio (N_1, N_2, N_3)	7:5:1
Magnetizing inductance (L_m)	300 μH
Total leakage inductances (L_1, L_2', L_3')	16, 15 and 0.28 μH
Secondary side output capacitor (C_2)	86 μF
Tertiary output capacitor (C_3)	47 μF
Switching frequency (f_{sw})	100 kHz

in this example is defined as the sum of squares of the winding current RMS values - all reflected to one side, i.e., $F_{cost}(x) = F_{RMS\ currents}(\delta_j, \varphi_j, f_{sw})$.

Figure 4.5 illustrates the algorithm used to identify the optimal PWM strategy for a particular operating condition. The core idea behind this search for an optimal modulation technique is to employ a PWM method that minimizes control degrees of freedom (i.e., reduces computational complexity) while ensuring that $F_{cost}(x)$ remains within 1.04 times the value obtained with PPS-based modulation i.e., within $\pm 4\%$ band from the best-case optimization. Under these conditions, we search for the optimal PWM technique across a wide range of m_2, m_3, P_2, and P_3 values, as depicted in Fig. 4.6.

The data presented in Fig. 4.6 covers a broad range of m_2 (from 0.7 to 1.3) and P_2 (from 50 to 400 W) values under nine combinations of m_3 and P_3. These combinations reflect low, unity, and high gain ($m_3 = 0.8, 1$, and 1.2) as well as light, mid, and high load ($P_3 = 50, 200$, and 400 W) at port-3. The percentage reduction in the computed objective function $F_{cost}(x)$ under the optimal PWM strategy, compared to the basic DPS modulation, is also shown in a contour plot. This plot illustrates the benefits of the optimized PWM modulation technique over conventional DPS-based control for specific operating zones. Key observations from this plot include:

- The basic DPS PWM strategy becomes optimal over a broader (m_2, P_2) range as output power increases from mid- to full-load at a particular port-3 voltage gain.

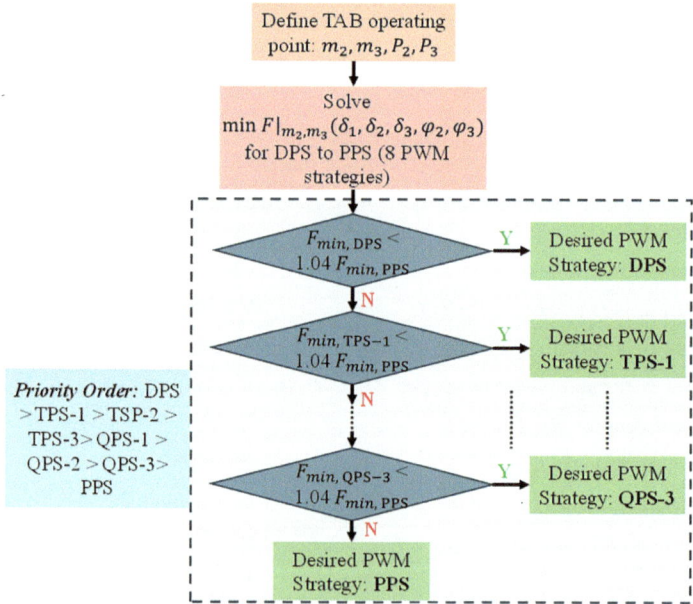

Fig. 4.5 Flowchart to achieve the optimum PWM strategy for any operating condition in a TAB

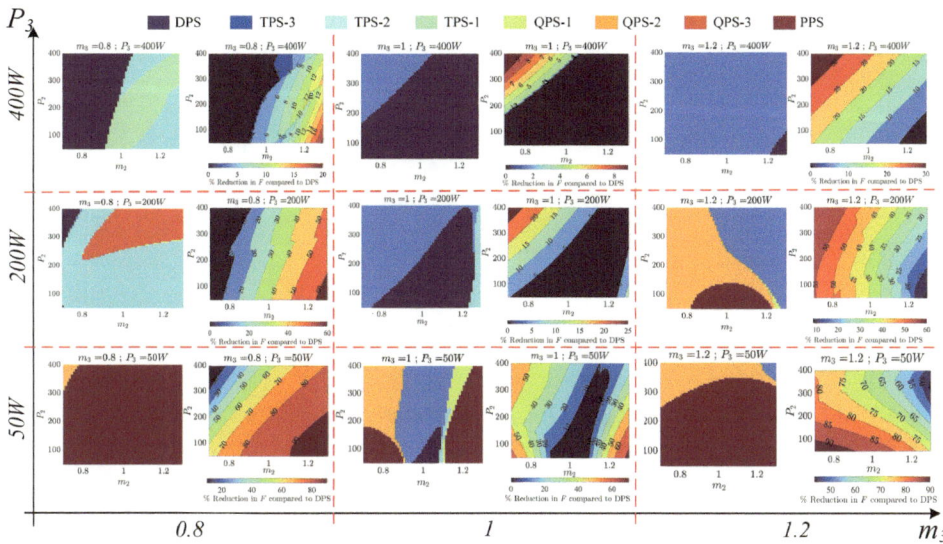

Fig. 4.6 Applicability of the PWM strategies in controlling TAB depending on the power and voltage levels

- As m_3 deviates from unity, the involvement of a higher-order PWM technique, such as Triple Phase-Shift (TPS), becomes necessary to maintain high load at port-3.
- At $m_3 = 1$, as P_3 decreases, TPS, Quad Phase-Shift (QPS), and PPS, depending on the (m_2, P_2) condition, emerge as optimal candidates for reducing conduction loss.
- For light load at port-3 i.e., $P_3 = 50W$, the utilization of the PPS becomes increasingly critical with the shift in m_3 gain.

Overall, both composite and optimized PWM control techniques become more advantageous as the converter approaches lighter loads and non-unity gain operation at any output port.

Figure 4.7a shows the variation in $F_{cost}(x)$ achieved by different PWM strategies under wide gain and light load (50 W) conditions at port-3 for an operating point of (m_2, P_2) = (1, 174 W). Similarly, Fig. 4.7b illustrates the variation in $F_{cost}(x)$ for different PWM strategies under wide gain and medium load (200 W) conditions at port-3 for an operating point of (m_3, P_3) = (1, 18 W). These plots clearly demonstrate that at non-unity voltage gain conditions, the application of PPS is essential for optimizing conduction loss in the TAB system. Furthermore, building on the initial case study, the optimization of control variables in a TAB converter can be approached using different cost functions $F_{cost}(x)$. To illustrate this, we also analyze the optimization results when the objective function is related to the switching loss in the MOSFETs rather than conduction loss, hence, $F_{cost}(x) = F_{sw_{loss}}(x)$. Moving forward, the multi-variable, multi-constrained TAB

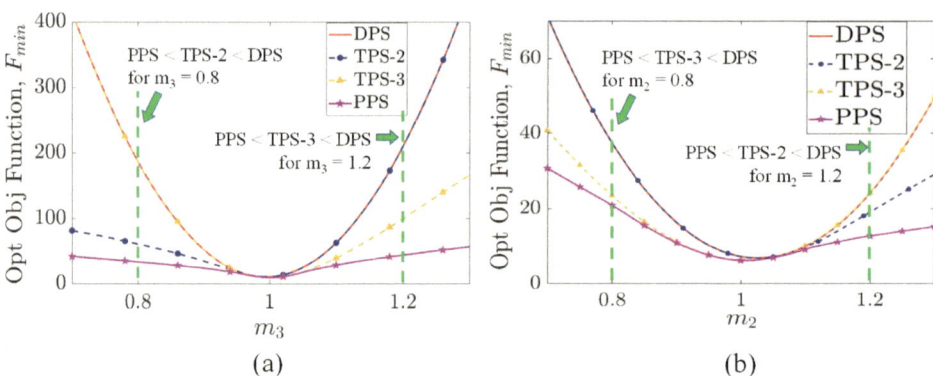

Fig. 4.7 **a** Optimum PWM technique for a wide variation in m_3, for $(P_3, m_2, P_2) = (50 \text{ W}, 1, 174 \text{ W})$. **b** Optimum PWM technique for a wide variation in m_2, for $(P_2, m_3, P_3) = (200 \text{ W}, 1, 18 \text{ W})$

switching loss optimization problem is tackled by applying various PWM strategies (from DPS to PPS) in the TAB converter, as exemplified previously. Figure 4.8 shows the variation of the optimized objective function $F_{cost}(x)$ for a wide range of port-3 voltage gain while maintaining unity gain at port-2, under different PWM strategies: DPS, TPS, and PPS. It is clear that for both over and under unity gain conditions, PPS achieves a minimized objective function compared to other PWM control schemes, which is also established through the conduction loss optimization results.

Figure 4.9 compares the variation of total switching and conduction losses at the switches in the developed TAB converter under different PWM control schemes for (a) over-unity gain ($m_3 = 1.2$) and (b) under-unity gain ($m_3 = 0.8$) conditions. The results indicate that the optimal PPS-based control substantially minimizes both switching and conduction losses at lighter loads compared to DPS or TPS-based control.

Fig. 4.8 Optimized objective function plot with different control techniques for a variation in m₃

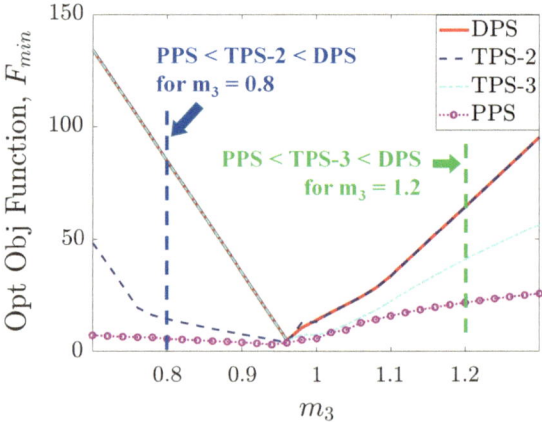

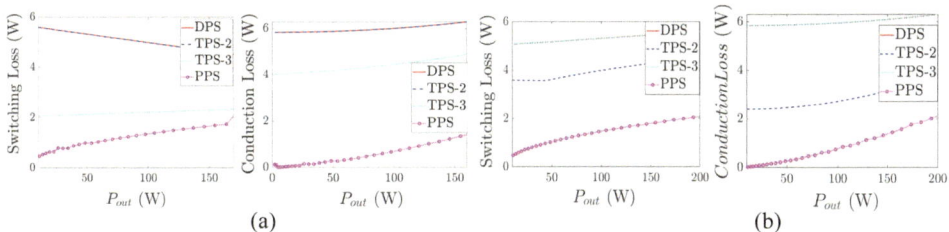

Fig. 4.9 Variation of the switching and conduction losses with a sweep in light load output power for **a** 160 to 114 and 27.4 V conversion and **b** 160 to 114 and 18.3 V conversion under different PWM schemes

Additionally, the effect of the proposed switching loss minimization algorithm on the soft-switching range of the converter is analyzed and depicted in Fig. 4.10. The ZVS conditions identified in Chap. 1 are used here to predict the soft-switching capability of any PWM strategy under application. It is noteworthy that while the PPS logic significantly improves system efficiency over basic DPS control, it also extends the ZVS range of the converter's switching legs across a wide voltage gain and load range.

4.3   Modulation Optimization in a DC-DC-AC TAB Converter

With the increased usage of the household level grid-tied solar-based energy resources, the high frequency link (HFL) single phase inverter topologies are gaining more popularity [16]. The recently developed dual-active-bridge (DAB) based HFL inverters [17–20] employ a single stage power conversion architecture that circumvented the requirement for the electrolytic intermediate dc link capacitor in traditional two-staged microinverters [21]; thus, producing a more reliable product with longer lifetime. The operation and modulation optimization of such DC-AC DAB are highlighted in Chap. 3. In addition to this, recent improvement in energy storage technology has substantially lowered the cost of engaging a battery storage system with the microinverter system. Employing an energy storage device enhances the power quality of the renewable energy system by improving its dispatchability, overall reliability and performance [22]. Besides, any renewable energy systems need to interface several energy sources such as PV arrays, wind turbines, fuel cells with the load or the ac grid along with battery backup. In a fuel cell-based system, the fuel cell generator will shut down or fail if the load demand exceeds its capacity. Consequently, the current demand must always be kept within the limits of the available current. Therefore, an additional port for bidirectional energy storage is needed to absorb the power difference. In systems where bidirectional power flow is required between the DC source, DC energy storage, and the AC grid, a three-port converter like the DC-DC-AC TAB converter is of particular interest because of its beneficial features.

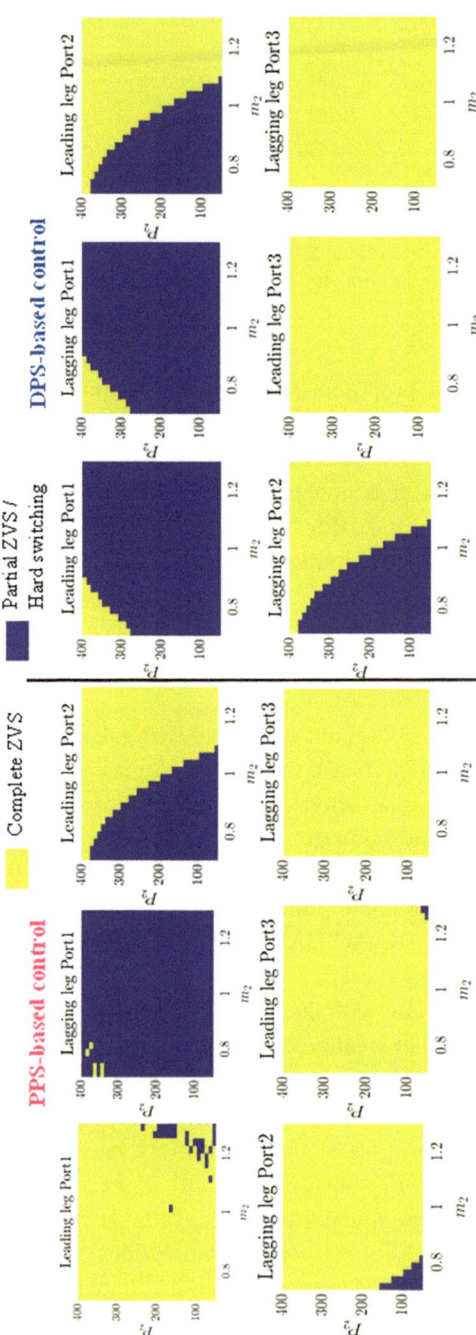

Fig. 4.10 Identification of ZVS zones based on PPS and DPS based TAB control for wide variation in voltage gain and power demand at port-2 while keeping the port-3 voltage gain at 1.2 at 50 W load

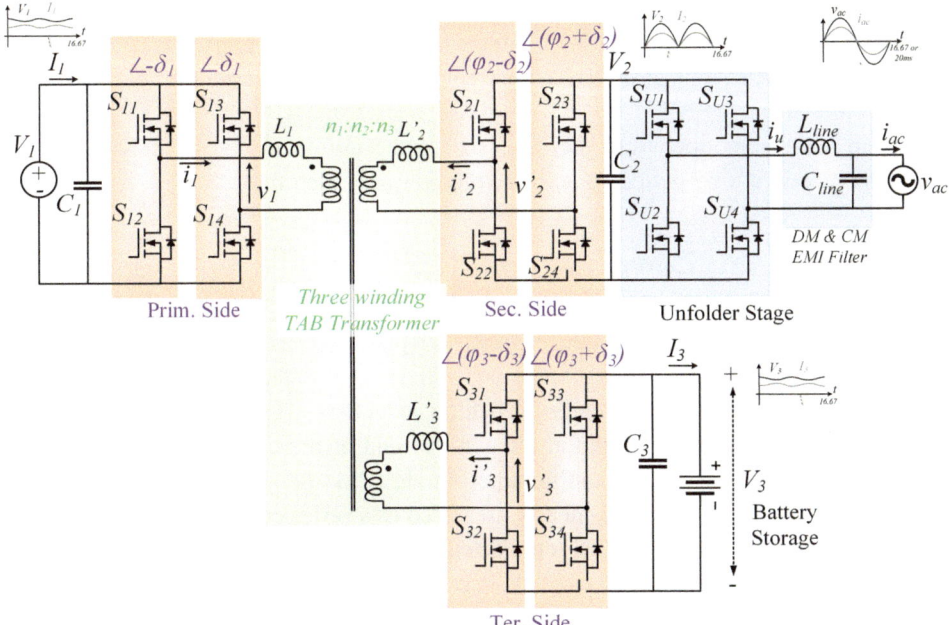

Fig. 4.11 Circuit Topology of the dc-dc-ac TAB Converter connecting single phase ac grid with the solar PV and an energy storage battery

Figure 4.11 depicts the schematic of the single-phase, single-stage, bidirectional and isolated TAB based dc-dc-ac converter topology. The ac port of the converter follows the dc-dc TAB stage by an unfolder and EMI filter network. Following the input source, the dc-dc TAB power conversion stage consists of three active bridges (ABs) coupled by a HF three-winding transformer having a turns ratio of $n_1 : n_2 : n_3$ and three series connected inductors that can often be realized by the leakage inductances of the transformer. The power transfer between the ABs is controlled by modulating the HF switching node voltages' (v_1, v_2' and v_3') duty cycles (δ_1, δ_2, and δ_3) and their relative phase differences (φ_2 and φ_3) that results from introducing phase shifts between the gating signals of the AB switches. The range of these control parameters can be given as: $\delta_k \rightarrow \left\{0, \frac{\pi}{2}\right\}$ and $\varphi_k \rightarrow \left\{-\frac{\pi}{2}, \frac{\pi}{2}\right\}$, where $k \in \{1, 2, 3\}$. The possible PWM modulation techniques in a dc-dc-ac TAB converter are similar to the dc-dc TAB and can be categorized into 4 groups based on control degrees of freedom and their combinations: Dual Phase Shift (DPS) $\{\varphi_2, \varphi_3\}$, Triple Phase Shift (TPS)$\{\varphi_2, \varphi_3$ and one of $\delta_k\}$, Quad Phase Shift (QPS)$\{\varphi_2, \varphi_3$ and two of $\delta_k\}$ and Penta Phase Shift (PPS)$\{\varphi_2, \varphi_3$ and all of $\delta_k\}$.

Thus, by independently varying all the phase shift (φ_k) and duty cycle (δ_k) control variables, which is commonly called as PPS modulation, the secondary side TAB output voltage $V_2(t)$ is regulated in a shape of a dc voltage that pulsates at twice the ac line frequency f_L (i.e., $V_2(t) = |v_{ac}(t)| = \left|\widehat{V_{ac}}sin(\omega_L t)\right|$, where $\omega_L = 2\pi f_L$) while the 3rd TAB port output is maintained as a stiff dc voltage of V_3. The ac link capacitor C_2 connected at the ac port of TAB sinks in the HF ac current ripple and a dc current propagates towards the unfolder stage.

During converter operation, the state st_U of the unfolder stage switches (S_{U1}–S_{U4}) toggles twice in a time period of ac line voltage $v_{ac}(t)$ such that,

$$st_{U1,U4,\overline{U2},\overline{U3}} = \begin{cases} 1, if\ v_{ac}(t) > 0 \\ 0, if\ v_{ac}(t) < 0 \end{cases} \tag{4.25}$$

Following this continuous state change, the pulsating dc voltage $V_2(t)$ is unfolded and forms a sinusoidally varying ac output voltage $v_{ac}(t)$. The voltage drops across the differential mode (DM) output stage EMI filter inductors are neglected in a steady state. The EMI filter stage is connected at the extreme end of the converter that interfaces the grid and is comprised of DM and common-mode (CM) filter components. All the DM mode filter impedances are equivalently presented as a series connected line inductor L_{line} on the ac line and a capacitor C_{line} connected across the line-neutral terminals.

The mathematical modeling of a dc-dc-ac TAB converter is exactly similar to the modeling of a dc-dc TAB performed earlier. The only difference is that for the dc-dc-ac TAB, one of the TAB port outputs is regulated as a rectified sinusoid voltage ($V_2(t) = |v_{ac}(t)| = \left|\widehat{V_{ac}}sin(\omega_L t)\right|$). In a TAB based dc-dc-ac converter, the major portion of the switching network losses include the conduction and switching losses that occur due to the high frequency switching of the ABs that are directly dependent on the TAB transformer winding currents RMS and their instantaneous values, as we observed for the case of a dc-dc TAB. The unfolder stage of the converter switches at ac line frequency, thus, causes only conduction loss with negligible switching loss. Because its action is aligned with the ac line voltage and cannot be modulated like the HF switching stage, its loss cannot be optimized with PWM modulation. Hence, the task of optimizing the dc-dc-ac TAB modulation solely concentrates on optimizing the losses incurred by the HF TAB stage that is same for any dc-dc or dc-dc-ac TAB converter. Due to the time varying voltage nature of the ac port of the TAB under study, the optimized modulation variables need to be time varying as well. Thus, while optimizing the control variables we consider the ac port voltage as piecewise linear and solve the optimization problem for each voltage step.

The relation between the TAB transformer winding currents i'_k and the AB voltages v'_k and the respective line inductances L'_k can be deduced from the simplified Δ-equivalent model of the TAB switching network, from which the loss model of the HF TAB stage is deduced.

4.3.1 Loss Function Formulation

The analytical formulations of the conduction and switching loss components and their related cost functions are discussed below.

(i) Conduction Loss

The conduction loss in a switching cycle in the HF switches connected in the three ABs depends on the individual port's winding current RMS and the on-state drain-source resistance $R_{DS(ON)}$ of the respective AB MOSFETs:

$$F_{cond\_loss}\left(\delta_k, \varphi_k, V_k\right) = \sum_{k=1}^{2} i_{k,RMS}^2 R_{DS(ON)_k} \qquad (4.26)$$

where, $i_{k,RMS}$ is the kth transformer winding current RMS, shown in from (4.2)–(4.5), and $R_{DS(ON)_k}$ is the normalized on-state resistance of the kth AB switches with respect to the primary bridge.

(ii) Switching Loss

The switching loss in a TAB converter majorly depends on the switch currents and their blocking voltages during the switching transients and the switching frequency of operation and its model is derived in case of a dc-dc TAB, and is represented $F_{sw\_loss\_total}\left(\delta_j, \varphi_j, f_{sw}\right)$.

$$F_{sw\_loss\_total}\left(\delta_j, \varphi_j, f_{sw}\right) = F_{sw\,loss}\left(\delta_j, \varphi_j, f_{sw}\right) + F_{Coss}\left(\delta_j, \varphi_j, f_{sw}\right) \qquad (4.27)$$

The ZVS conditions for a TAB switching network depend on the switching instant currents, dc link voltages and the device's parasitic body capacitors. In case of the dc-dc TAB, we assume the parasitic body capacitors of the switching MOSFETs (C_{OSS}) that determines the ZVS condition, to be constant throughout the converter operation as the dc link voltage is nearly constant throughout the steady state converter operation.

However, this capacitance increases substantially with a decreasing drain-source voltage (v_{DS}) across the device. In our dc-dc-ac converter application, for the secondary side (ac port) H-bridge, the v_{DS} of the MOSFETS varies in a wide range of voltage from 0 V to $\widehat{V_{ac}}$. Thus, the condition for the secondary side AB switches' ZVS also becomes very stringent near zero output ac line voltage due to a higher C_{OSS} value. Therefore, to precisely synthesize the ZVS constraints, the C_{OSS} of the HF TAB MOSFETs are modeled as a closed form function of the v_{DS}, using regression model-based curve fitting approach, as presented in (4.28).

Table 4.4 Switching current informed ZVS constraints for kth active bridge ($k = 1, 2,$ or 3; $\varphi_1 = 0$)

Case	Turn-on switches	ZVS condition	Switching time (τ)
$\begin{cases} \delta_k = 0 \\ i_k(\tau) > 0 \end{cases}$	S_{k1}, S_{k4}	$\frac{1}{2} L i_k(\tau)^2 \geq 2\left(K_2 V_k + K_1 \sqrt{V_k}\right) v_{eq,k}(\tau)$	$\varphi_k + \pi$
$\begin{cases} \delta_k = 0 \\ i_k(\tau) < 0 \end{cases}$	S_{k2}, S_{k3}	$\frac{1}{2} L i_k(\tau)^2 \geq -2\left(K_2 V_k + K_1 \sqrt{V_k}\right) v_{eq,k}(\tau)$	φ_k
$\begin{cases} \delta_k \neq 0 \\ i_k(\tau) > 0 \end{cases}$	S_{k2}	$\frac{1}{2} L i_k(\tau)^2 \geq \left(2 v_{eq,k}(\tau) V_k - V_k^2\right)\left(K_1 + \frac{K_2}{\sqrt{V_k}}\right)$	$\varphi_k + \pi - \delta_k$
$\begin{cases} \delta_k \neq 0 \\ i_k(\tau) < 0 \end{cases}$	S_{k1}	$\frac{1}{2} L i_k(\tau)^2 \geq \left(-2 v_{eq,k}(\tau) V_k - V_k^2\right)\left(K_1 + \frac{K_2}{\sqrt{V_k}}\right)$	$\varphi_k - \delta_k$
$\begin{cases} \delta_k \neq 0 \\ i_k(\tau) > 0 \end{cases}$	S_{k3}	$\frac{1}{2} L i_k(\tau)^2 \geq \left(2 v_{eq,k}(\tau) V_k + V_k^2\right)\left(K_1 + \frac{K_2}{\sqrt{V_k}}\right)$	$\varphi_k + \pi + \delta_k$
$\begin{cases} \delta_k \neq 0 \\ i_k(\tau) < 0 \end{cases}$	S_{k4}	$\frac{1}{2} L i_k(\tau)^2 \geq \left(-2 v_{eq,k}(\tau) V_k + V_k^2\right)\left(K_1 + \frac{K_2}{\sqrt{V_k}}\right)$	$\varphi_k + \delta_k$

$$C_{OSS}(v_{DS}) = \frac{K_1}{\sqrt{v_{DS}}} + K_2 \tag{4.28}$$

The coefficients K_1 and K_2 depend on the choice of the power devices and can be calculated from the C_{OSS} versus v_{DS} plot as per the corresponding MOSFET datasheets. Keeping such a C_{OSS} model in mind, the possible cases of TAB switching transients and their respective ZVS requirements are reevaluated and presented in the Table 4.4 as a summary.

4.3.2   DC-DC-AC TAB Control Optimization Problem Formulation

Upon the completion of analytical modeling of the TAB switching network losses, a loss optimization algorithm can be synthesized that results in the optimal generalized hybrid-phase-duty-frequency based PPS modulation scheme for the TAB based dc-dc-ac converter. The optimization problem can be formulated similarly to how we approached it for the dc-dc TAB. For a varying output voltage i.e., rectified AC sinusoid, such problem can be solved multiple times to deduce the optimal control variables ($x_{opt} = \{\delta_1, \delta_2, \delta_3, \varphi_2, \varphi_3, f_{sw}\}$) across the ac line voltage. The formulation of the cost function $f_{cost}(x)$ for optimization for a dc-dc-ac TAB is same as the dc-dc TAB converter and is presented in (4.24). The non-linear equality constraints in case of a dc-dc-ac TAB optimization are dictated by: (a) the desired amount of instantaneous power to be supplied to the secondary side ac output during any particular ac cycle phase angle and (b) power transferred to the other dc port side (tertiary side). If the port-2 and port-3 are treated as the ac and dc output ports, respectively and the instantaneous power demand of these ports are $p_2(t)$, and P_3, respectively, the power flow constraints can be formed by equating each of them with the GHA model-derived port power expressions, as given in (4.29) and (4.30).

$$
\begin{aligned}
p_2(t) &= \frac{\left|\widehat{V_{ac}} \sin(\omega_L t)\right|^2}{R_{load}} \\
&= \frac{8P_{base}}{\pi^2} \sum_{k=1}^{2m+1} \frac{1}{k^3}\left[\frac{3m_2 L_1}{L_{12}} d_{k1} d_{k2} \sin(k\varphi_2) + \frac{3m_2 m_3 L_1}{L_{23}} d_{k2} d_{k3} \sin\{k(\varphi_2 - \varphi_3)\}\right]
\end{aligned}
$$

(4.29)

$$
P_3 = \frac{8P_{base}}{\pi^2} \sum_{k=1}^{2m+1} \frac{1}{k^3}\left[\frac{3m_3 L_1}{L_{13}} d_{k1} d_{k3} \sin(k\varphi_3) + \frac{3m_2 m_3 L_1}{L_{23}} d_{k2} d_{k3} \sin\{k(\varphi_3 - \varphi_2)\}\right]
$$

(4.30)

where $P_{base} = \frac{V_1^2}{2\pi f_{sw} L_1}$, $d_{k1} = \cos(k\delta_1)$ and $d_{k2} = \cos(k\delta_2)$, R_{load} is the resistive load connected at the ac side.

The other constraints are the functions imposing limitation on the decision variables. These types of functions (4.31) set the upper and lower bounds to the x so that the solution stays inside the feasible zone.

$$
0 \leq \delta_k \leq \frac{\pi}{2}; -\frac{\pi}{2} \leq \varphi_k \leq \frac{\pi}{2}; f_{sw,min} \leq f_{sw} \leq f_{sw,max}.
$$

(4.31)

For a given dc-dc-ac TAB converter parameter, for a wide range in output load and for each phase angle degree of the 60 Hz ac line cycle (i.e., $1°$, $2°$, ... $180°$), the loss optimal generalized PPS modulation control variables (δ_k, φ_k, f_{sw}) can be derived by minimizing the total switching network loss equivalent cost function $f_{cost}(x)$ while maintaining the constraints such as load power, upper and lower bounds of the decision variables.

4.3.3 Case Study: PWM Strategy Optimization for a DC-DC-AC TAB

In this case study, we delve into optimizing the control variables of a fixed frequency-based PPS PWM strategy for a three-port (TAB) DC-DC-AC converter. The converter under study represents a 500 W PV microinverter integrated with battery storage. It has an input side solar PV with its nominal maximum power point (MPP) at 40 Vdc (V_1)

and 500 W of maximum output power (P_1). Further, the ac output port of the converter is regulated at 120 Vac RMS voltage with 60 Hz line frequency supplying 250 W of maximum average active power, while the 250 W rated tertiary port output of the converter is connected to a dc battery with 28 V. The major circuit parameters of the example converter are highlighted in Table 4.5.

While optimizing the control variables using the optimization problem defined in last section, the cost function $f_{cost}(x)$ is made as the total loss in the HF TAB switching network including the conduction as well as switching loss, i.e., $F_{cond_{loss}}(x) +$ $F_{sw\_loss\_total}(x)$. Further, the f_{sw} of the converter is kept constant at 100 kHz for this analysis. Now, the optimization algorithm with the finalized constraint functions and design variables is solved using numerical optimization toolbox in MATLAB for the optimal decision variable set x_{opt} for each ac line phase angle (Φ) from 0° to 180° in 1° interval at a specific input dc voltage $V_1 = 40V$, output dc voltage, $V_{3'} = 28V$ and a wide range of output load conditions: $P_{ac,RMS} \in [0W, 250W]$, $P_3 \in [0W, 250W]$ and the x_{opt} array is stored in a $M \times N$ matrix. Here, N = 8, representing the number of columns housing

Table 4.5 Example DC-DC-AC TAB converter circuit components

Circuit parameters	Values
Input DC voltage (V_1)	40 VDC
Primary H-bridge switches	EPC2033 150 V 48 A 7 mΩ GaN FET
Primary side DC link capacitor (C_1)	9.9 mF/100 V
Secondary output port voltage (V_2)	120 Vac
Secondary output port rated power ($P_{2,avg}$)	250 W
Secondary HF H-bridge switches	GS66516T-MR 650 V 60 A 32 mΩ GaN FET
Secondary side pulsating DC link capacitor (C_2)	2 μF/500 V
Secondary side unfolder bridge switches	C3M0120065D 650 V 22 A SiC MOSFET
Tertiary output port voltage (V_3)	28 VDC
Tertiary output port rated power (P_3)	250 W
Tertiary H-bridge switches	EPC2020 60 V 90 A 2.2 mΩ GaN FET
Tertiary side DC link capacitor (C_3)	6.6 mF/80 V
Transformer turns ratio (N$_1$, N$_2$, N$_3$)	4:12:3
Magnetizing inductance (L_m)	280 μH
Total leakage inductances (L_1, L_2', L_3')	1.65, 27 and 0.93 μH
Switching frequency (f_{sw})	100 kHz
DM filter inductance (L_{line})	80 μH
DM filter capacitance (C$_{line}$)	2 μF/310 Vac

each input variables to the optimization problem Φ, P_2(instantaneous power at ac port $= 2P_{ac,RMS} \sin^2 \Phi$), P_3, against the optimal control parameter outputs $\delta_1, \delta_2, \delta_3, \varphi_2, \varphi_3$. M depicts the number of distinct datasets that can be generated by the possible combinations of the optimization inputs, Φ, P_2, P_3, i.e., $180 \times 10 \times 10 = 18,000$. The synthesized optimization routine is also described in the form of a flowchart in Fig. 4.12. A series of results from this optimization over a half ac cycle is depicted in Fig. 4.13, where the $P_{2,RMS}$ and P_3 are fixed at 250 W and 50 W, respectively. Here, the optimal TAB control parameters along with the resulting losses at the switches over a half ac line cycle are presented for a PPS and traditional DPS modulation strategy.

The results demonstrate that applying AB duty variables under PPS modulation results in a 16.7% lower power loss at the high-frequency switches during peak AC power delivery, due to reduced winding current peaks and their RMS values. Additionally, the optimal PPS scheme significantly reduces conduction losses (~ 74%) near the zero-crossing zone of the AC voltage output and extends the ZVS range across a broader AC line cycle, ensuring lower switching losses in the active devices.

The ZVS range of the TAB converter is further investigated under light and heavy load conditions with PPS modulation to understand its impact on the ZVS boundary. This investigation involves analyzing the ZVS feasibility for the six switching legs arranged in pairs within three active bridge configurations under two distinct loading scenarios: a 10% load (25 W to both secondary and tertiary ports) and a full 100% load (250 W to each output port from the primary DC port). Figure 4.14 graphically illustrates these outcomes.

The binary data based on the positions of the switching legs help discern the potential for ZVS within individual legs. For example, a code like '11 01 10' indicates both primary bridge legs under full-ZVS, the leading leg of the secondary bridge in hard-switching mode, the lagging leg of the secondary bridge in soft-switching mode, the leading leg of the tertiary bridge in soft-switching mode, and the lagging leg of the tertiary bridge in hard-switching mode. The crucial findings reveal that the instantaneous power at the AC secondary port in the DC-AC-DC converter, corresponding to the AC voltage phase angle, determines the ZVS possibilities. When the AC phase angle is close to 0 or $\pi/2$, or under light load conditions at the AC port, both switching legs of the secondary bridge miss soft-switching. A comparative analysis between light and heavy loads shows a higher prevalence of soft-switching for an extended range of AC voltage angles during heavy load operation. Despite the negative impact of light load operations on ZVS conditions, the introduction of the proposed PPS technique expands the range of soft-switching across all loading conditions, a trend highlighted in Fig. 4.13 as well.

In conclusion, the loss-optimal duty cycle tracking with PPS modulation is essential for a DC-DC-AC TAB converter, where the AC port output voltage varies with twice line frequency i.e., 120 Hz alongside a wide converter load range. This case study underscores the necessity of tailored PWM strategies to optimize different aspects of TAB converter performance, ensuring efficiency and reliability across diverse operating conditions.

Fig. 4.12 Complete
mathematical framework to
formulate the polynomial fitted
optimal duty cycle expressions
for implementation of the
minimum loss tracking
algorithm in a TAB dc-dc-ac
converter

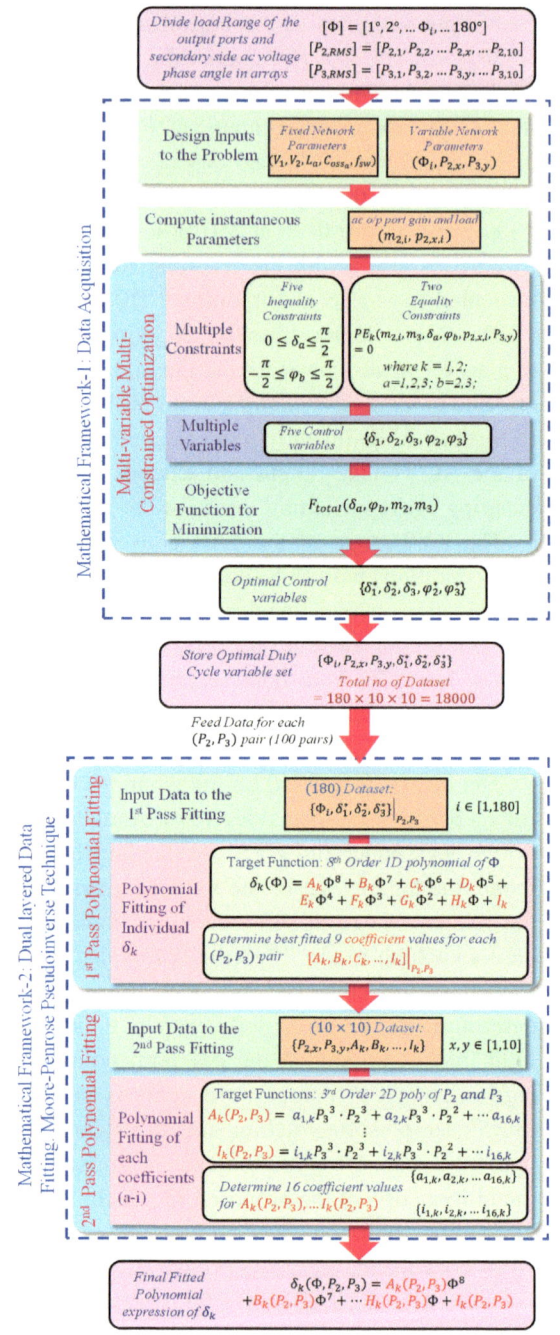

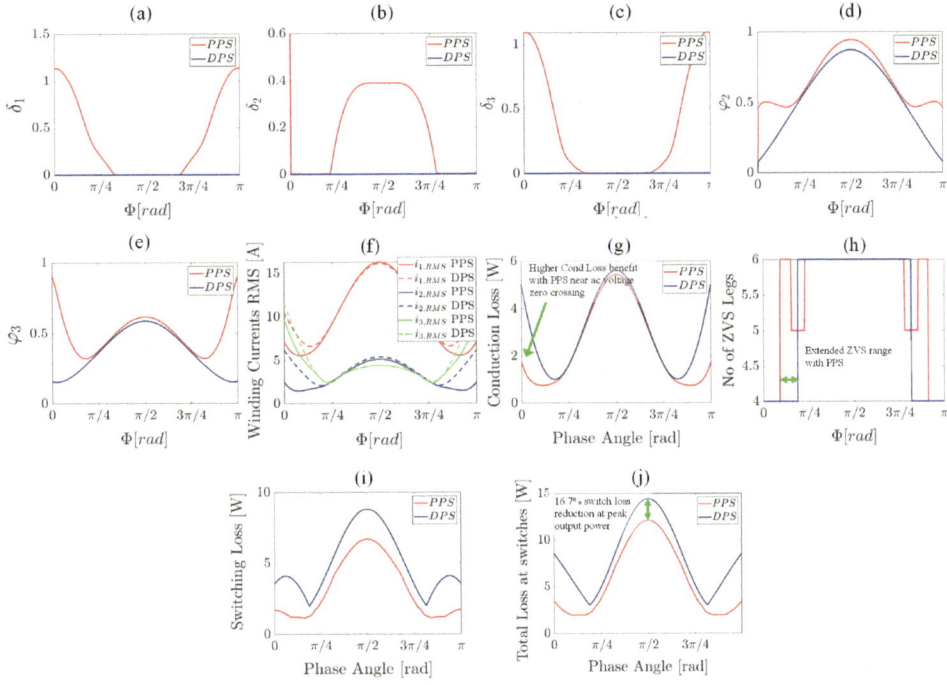

Fig. 4.13 Switching network Loss optimized TAB based dc-dc-ac converter operation at 40 to 120 Vac and 28 Vdc power conversion mode at $P_2 = 250$ W and $P_3 = 50$ W comparing optimal PPS modulation and conventional DPS modulation strategy over half ac line period: **a** δ_1; **b** δ_2; **c** δ_3; **d** φ_2; **e** φ_3; **f** transformer winding current RMSs; **g** conduction loss; **h** number of HF TAB legs undergoing ZVS; **i** switching loss; **j** total loss at the switches

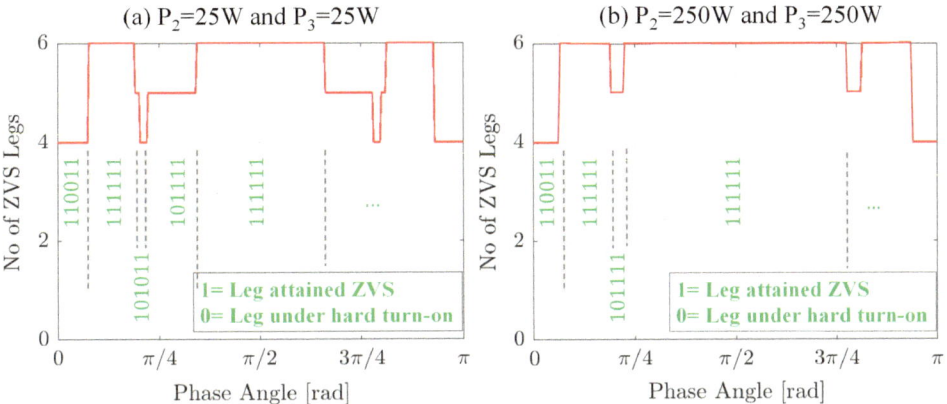

Fig. 4.14 Number of switching legs undergoing ZVS at **a** light load, or, $P_2 = 25$ W and $P_3 = 25$ W and **b** heavy load, or, $P_{ac,RMS} = 250$ W and $P_3 = 250$ W

4.4   Implementation of the Optimized Control in a TAB Converter Hardware

Upon the identification of the favorable as well as optimized PWM strategy for the TAB for specific port voltages and power levels, the optimum control variables are to be attained through an algorithm, realizable in the TAB controller. As mentioned earlier in this chapter, in the DPS PWM strategy, two degrees of freedom, i.e., φ_2 and φ_3 are the only available system control variables. For any dual-output multiple voltage electrical systems, two output port currents of the converter, i.e., I_2' and I_3', or the output voltages, V_2' and V_3' are the control objectives. Such two input (φ_2 and φ_3) and two output (I_2', I_3', or V_2' and V_3') control system can be realized through two separate control loops. However, for higher order PWM strategies with inclusion of the duty cycle variables (δ_1, δ_2, and δ_3) in the control system, hardware implementation becomes a challenge. Based on the operating load and DC port voltage levels, the optimized control variables, specifically the switching frequency and duty cycles of the bridge voltages need to be properly selected. In order to perform this task, two methods of digital controller based optimal modulation implementation approaches are discussed below.

4.4.1   Look-Up Table Based Approach

Many of the existing works on DAB and TAB converter use the look-up table-based approach during optimized modulation implementation [4]. A TAB with the optimized conduction loss target will need a specific optimal control variable set for any operating point defined by the voltage gain and load conditions $\{m_2, m_3, P_2, P_3\}$ of the respective output ports. To generate a look up table comprising of the duty cycle variables for each of the operating conditions will need multiple 3D arrays or 4D arrays to be computed first and then implemented in the DSP. This involves an extensive computational effort and requires a large memory in the controller. Moreover, the formulation of the lookup table while incorporating such implementation strategy solely depends on the estimated or measured circuit parameters of the lossless TAB model such as leakage/line and magnetizing inductances. In hardware, these values fluctuate due to changes in load, aging effects, and varying power losses, making accurate estimation a difficult task. As a result, often during practical implementation, these pre-calculated optimum control values do not lead to the least conduction loss operation.

4.4.2 Multivariable Polynomial Fitted Model Development and Implementation

As look-up table (LUT) based deployment of the offline optimized control parameters is not scalable for MABs with three or more number of ports, due to its extremely large array size and memory allocation challenges, another alternative approach for implementing the dynamically adaptive optimized control variables is 'Polynomial Regression based model deployment'. In this method, we mathematically obtain the polynomial regression-based models for the loss optimal duty cycle control variables of a TAB converter that can be easily implemented inside the DSP to achieve the desired control variables based on the running load and input/output voltage conditions. If the model is precisely fitted using offline optimization data collected from all possible operating data space of the converter loads and dc/ac port voltages, such models can be effectively used to dynamically change the optimal control variables on the fly based on sensed port voltages and load currents.

There are many possible ways to construct such polynomial regression models; however, all the methods follow the two steps: data collection and model development using data fitting. One effective way to collect model data and mathematically construct the model for a dc-dc-ac TAB converter is presented below.

- **Data Collection:** In the dc-dc-ac TAB converter case study presented in Sect. 4.3.3, with the employment of a constant frequency PPS modulation, the optimal phase-duty parameters were synthesized by solving the multi-variable multi-constrained loss optimization problem for all possible load and voltage gain conditions. Thus, with a fixed input voltage of 40 V, fixed tertiary side dc voltage of 28 V, a piecewise linear output voltage at the 120 V ac port for every $1°$ phase angle, and 10 distinct load values at each output ports, total of 18,000 (=180 × 10 × 10) distinct load and voltage gain conditions are generated. For each of these cases, the optimal control parameters for the PPS scheme $x_{opt} = \{\delta_1, \delta_2, \delta_3, \varphi_2, \varphi_3\}$ are synthesized. Now these 18,000 data sets will be used to construct the polynomial models of the optimal duty variables as a function of the operating load and ac voltage phase angle or voltage gain. The phase shift variables φ_2, φ_3 are not modeled using such fitted functions as they are generated from the PI controller outputs to regulate the output load and voltages at desired level. If we derive all the variables of the control system including φ_2, φ_3 using polynomial fitted functions then there is a chance that the output voltage or current may be regulated at a lower value compared to their desired reference points because although we intend to generate the polynomial models of the control variables with least modeling error, there can still be some mismatch between the mathematical model of the converter and the actual hardware due to presence of circuit parasitics. To cater for this mismatch, the φ_2, φ_3 variables are obtained as outputs of the PI controllers that minimizes the error between the reference output voltage/current and their sensed values. The block diagram representing the closed loop control system for a TAB converter is showcased in Fig. 4.15.

- **Data Fitting:** While fitting the data to construct models for each duty variables $\delta_k (k = 1, 2, 3)$, a two-pass fitting approach is considered.

During the first pass fitting of the data, each of the optimal δ_k is targeted to fit into a one-dimensional 8th order polynomial function of Φ for each constant P_2, P_3 pair. There are 100 such possible pairs with 10 distinct $P_{2,RMS}$ [25 W, 50 W, … 250 W] and P_3 [25 W, 50 W, … 250 W] values: pair-1 [25 W, 25 W], pair-2 [25 W, 50 W], pair-3 [25 W, 75 W],… pair-100 [250 W, 250 W]). The optimal δ_k model per P_2, P_3 pair is given in (4.32) as a function of Φ alone.

$$\delta_k(\Phi) = A_k\Phi^8 + B_k\Phi^7 + C_k\Phi^6 + D_k\Phi^5 + E_k\Phi^4 + F_k\Phi^3 + G_k\Phi^2 + H_k\Phi + I_k \quad (4.32)$$

where, $A_k, B_k, \ldots I_k$ are constant coefficients of the fitted polynomial function, whose values are needed to be found.

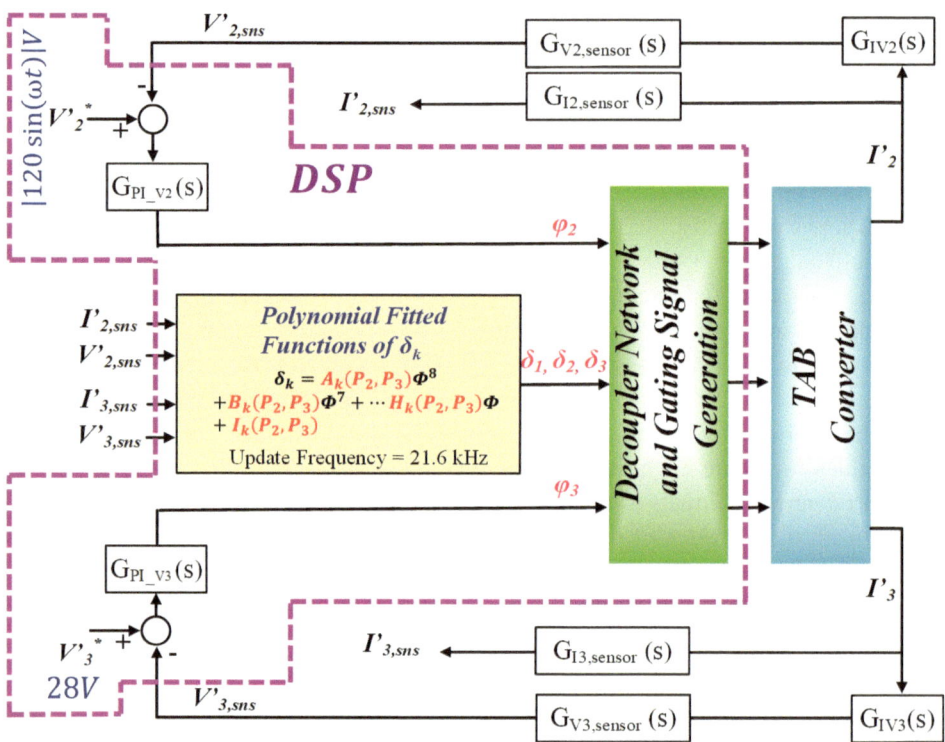

Fig. 4.15 Closed loop implementation of hybrid phase-duty oriented loss minimized TAB control incorporating the optimal duty tracking algorithm

For a particular power pair (among 100), 180 different values of the optimal δ_1 are there in the dataset that corresponds to 180 different phase angle Φ values. Thus, using (4.32) the fitted equations of δ_1 can be written as:

$$\delta_{1,1}(\Phi) = A_1 \Phi_1^8 + B_1 \Phi_1^7 + C_1 \Phi_1^6 + D_1 \Phi_1^5 + E_1 \Phi_1^4 + F_1 \Phi_1^3 + G_1 \Phi_1^2 + H_1 \Phi_1 + I_1;$$

$$\vdots$$

$$\delta_{1,180}(\Phi) = A_1 \Phi_{180}^8 + B_1 \Phi_{180}^7 + C_1 \Phi_{180}^6 + D_1 \Phi_{180}^5 + E_1 \Phi_{180}^4 + F_1 \Phi_{180}^3 + G_1 \Phi_{180}^2 + H_1 \Phi_{180} + I_1.$$

$$(4.33)$$

where $\Phi_1 = 1°,\dots \Phi_{180} = 180°$.

Further, the data fitting procedure for any δ_k can be represented in a matrix format as shown in (4.34).

$$\begin{bmatrix} \Phi_1^8 & \Phi_1^7 & \dots & 1 \\ \Phi_2^8 & \Phi_2^7 & \dots & 1 \\ \vdots & \vdots & \ddots & \vdots \\ \Phi_{180}^8 & \Phi_{180}^7 & \dots & 1 \end{bmatrix}_{180 \times 9} \cdot \begin{bmatrix} A_k \\ B_k \\ \vdots \\ I_k \end{bmatrix}_{1 \times 9} = \begin{bmatrix} \delta_{k,1} \\ \delta_{k,2} \\ \vdots \\ \delta_{k,180} \end{bmatrix}_{1 \times 180} \qquad (4.34)$$

$$\text{or,} \ A \cdot \vec{u} = \vec{f}, \qquad (4.35)$$

where $\vec{u}$ represents the coefficient vector $[A_k B_k \dots I_k]^T$ and $\vec{f} = [\delta_{k,1} \delta_{k,2} \dots \delta_{k,180}]^T$. In order to solve this problem by obtaining the best fitted $\vec{u}$ vector, Moore–Penrose Pseudoinverse algorithm is utilized [11, 23]. Using this algorithm, $\vec{u}$ can be obtained as

$$\vec{u} = A^+ \cdot \vec{f}, \qquad (4.36)$$

where A^+ is the Pseudoinverse of A. Further, A^+ can be written as, $A^+ = (A^*A)^{-1}A^*$, where A^* is the Hermitian transpose of A. Thus, using this method, the 9 best fitted coefficient values of $\vec{u}$, are determined for each of the possible 100 (P_2, P_3) pairs.

During the 1st pass polynomial fitting process of δ_k as a function of the ac phase angle Φ, a specific set of coefficient values, $\vec{u} = [A_k B_k \dots I_k]$ are obtained for each P_2, P_3 pair. However, $\vec{u}$ varies based on different P_2, P_3 pair combinations. Therefore, a second pass fitting process is necessary to formulate each of the coefficients $A_k, B_k,\dots I_k$ as individual functions of TAB output port load parameters, P_2, P_3. For this purpose, the collected data (total 100 discrete datasets in the form of $\{P_{2,x}, P_{3,y}, A_k, B_k, \dots I_k\}$) from the 1st pass fitting process, works as the input to the second layered fitting algorithm. During this stage, each of the δ_k coefficients are targeted to fit into a 3rd order 2-D polynomial of P_2 and P_3, as portrayed in (4.37).

$$A_k(P_2, P_3) = a_{1,k} P_3^3 \cdot P_2^3 + a_{2,k} P_3^3 \cdot P_2^2 + \cdots a_{16,k}$$
$$\vdots$$
$$I_k(P_2, P_3) = i_{1,k} P_3^3 \cdot P_2^3 + i_{2,k} P_3^3 \cdot P_2^2 + \cdots i_{16,k}$$

(4.37)

where, $a_{1,k}, a_{2,k}, \ldots a_{16,k}$ and $i_{1,k}, i_{2,k}, \ldots i_{16,k}$ are the polynomial coefficients of the fitted $A_k(P_2, P_3)$ and $I_k(P_2, P_3)$ functions.

Furthermore, fitting the total 100 data for coefficient 'A_k' into (4.37) can be represented in a matrix equation format that is highlighted in (4.38).

$$\begin{bmatrix} P_{3,1}{}^3 P_{2,1}{}^3 & P_{3,1}{}^3 P_{2,1}{}^2 & \cdots & 1 \\ P_{3,1}{}^3 P_{2,2}{}^3 & P_{3,1}{}^3 P_{2,2}{}^2 & \cdots & 1 \\ \vdots & \vdots & \ddots & \vdots \\ P_{3,10}{}^3 P_{2,10}{}^3 & P_{3,10}{}^3 P_{2,10}{}^2 & \cdots & 1 \end{bmatrix}_{100 \times 25} \begin{bmatrix} a_{1,k} \\ a_{2,k} \\ \vdots \\ a_{16,k} \end{bmatrix}_{1 \times 16} = \begin{bmatrix} A_k(P_{3,1}, P_{2,1}) \\ A_k(P_{3,1}, P_{2,2}) \\ \vdots \\ A_k(P_{3,10}, P_{2,10}) \end{bmatrix}_{1 \times 100}$$

(4.38)

where $P_{2,1} = 25$ W, $P_{2,2} = 50$ W, … $P_{2,10} = 250$ W; $P_{3,1} = 25$ W, $P_{3,2} = 50$ W, … $P_{3,10} = 250$ W; $A_k(P_{3,1}, P_{2,1})$ is the A_k coefficient of $\delta_{k,1}$ corresponding to power pair-1 $(P_{3,1}, P_{2,1})$.

The equation in (4.38) can also be written as $B\vec{v} = \vec{g}$, where $\vec{v} = \begin{bmatrix} a_{1,k} a_{2,k} \cdots a_{16,k} \end{bmatrix}^T$ and $\vec{g} = \begin{bmatrix} A_k(P_{3,1}, P_{2,1}), A_k(P_{3,1}, P_{2,2}), \ldots A_k(P_{3,10}, P_{2,10}) \end{bmatrix}^T$. Now, similar to the 1st pass fitting, $\vec{v}$ can be determined as, $\vec{v} = B^+ \cdot \vec{g}$, where B^+ is the Pseudoinverse of B and can be obtained as, $B^+ = (B^* B)^{-1} B^*$; B^* is the Hermitian transpose of B. Following a similar approach, P_2, P_3 dependent polynomial functions for the other δ_k coefficients are also synthesized. The fitted results for coefficient '$A_1(P_2, P_3)$' and '$B_1(P_2, P_3)$' are presented in a form of surface plots in Fig. 4.16.

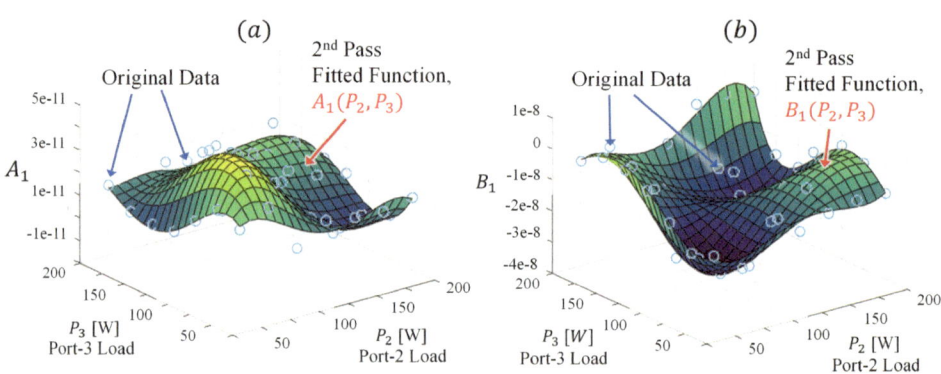

Fig. 4.16 Best fitted functions of the 2nd pass fitting coefficients of δ_k and their originally obtained values for wide variation in P_2 and P_3: **a** $A_1(P_2, P_3)$; **b** $B_1(P_2, P_3)$

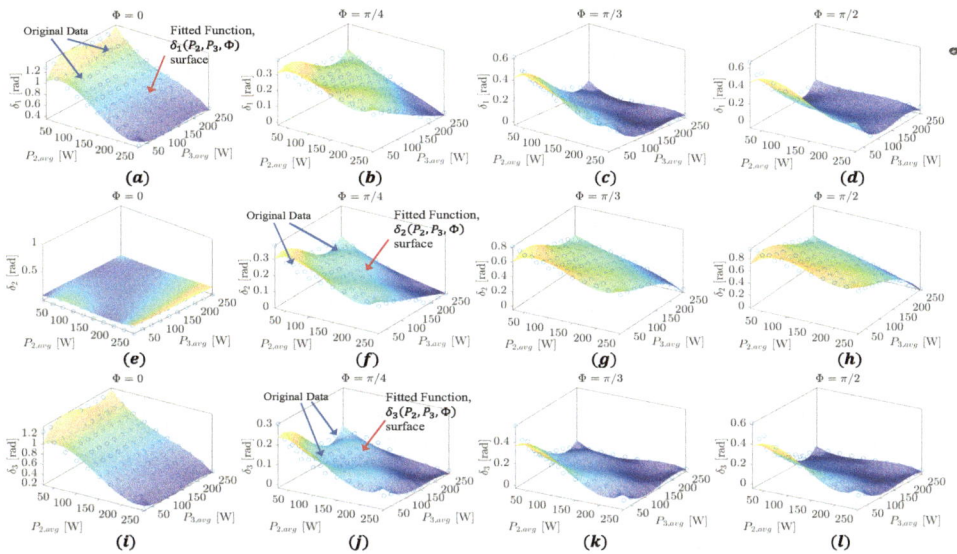

Fig. 4.17 Final polynomial fitted function plots of optimal δ_1, δ_2, and δ_3, and their originally obtained values δ_1^*, δ_2^*, and δ_3^* for wide variation in P_2 and P_3 under $\Phi = 0$, $\frac{\pi}{4}$, $\frac{\pi}{3}$, $\frac{\pi}{2}$

Therefore, upon completing this 2nd pass data fitting process, a complete polynomial fitted model of the optimal TAB duty cycle parameter δ_k is generated in (4.39) that is a continuous function of the converter's operating ac output phase angle and load levels.

$$\delta_k(\Phi, P_2, P_3) = A_k(P_2, P_3)\Phi^8 + B_k(P_2, P_3)\Phi^7 + \ldots H_k(P_2, P_3)\Phi + I_k(P_2, P_3) \quad (4.39)$$

Further, we verify the accuracy of the formulated optimal duty cycle model compared to the original input data set. Three sets of results, presented in Fig. 4.17, show the derived models for δ_1, δ_2, and δ_3 against their actual optimal values over a wide range of P_2 and P_3 variations. These variations are examined under four different v_{ac} phase angles ($\Phi = 0$, $\frac{\pi}{4}$, $\frac{\pi}{3}$, $\frac{\pi}{2}$). The fitting accuracy of the highly nonlinear optimal δ_1, δ_2, and δ_3 functions is evident from the results, with calculated variances of 4.2%, 3.8%, and 4.1%, respectively. This demonstrates that the polynomial-based regression models of the optimal duty variables effectively ensure loss-minimized operation of the TAB converter for any output voltage gain and load condition. These optimal duty models can be readily implemented within the TAB controller, thereby avoiding any memory allocation challenges. However, these polynomial models will impose a computational burden on the controller, which depends on the order of the polynomial functions.

Upon completing the synthesis of the optimal duty models, their implementation can be carried out in the closed-loop TAB converter control system, as depicted in Fig. 4.15. In this system, standalone PI controllers track the referenced output port voltages ($V_2'^*$

and $V_3'^*$) by dynamically adjusting the phase shift control parameters (φ_2, φ_3). Meanwhile, the duty-cycle parameters are updated by the synthesized optimal duty regression models to ensure minimized switching network loss operation. The PI controller outputs pass through a decoupler network to prevent undesirable coupling effects between the individual TAB output ports. Additionally, based on the sensed TAB output voltages and load currents, the optimal duty cycle tracking control block updates the operating bridge voltage duty values online within the controller.

It is important to note that the implemented polynomial functions of the optimal δ_k will impose some computational burden on the digital controller, particularly in terms of the program execution time needed for arithmetic operations, including a series of multiplications and additions. This computational time is directly correlated with the order of the modeled polynomial function. Higher-order polynomials can achieve greater model accuracy at the cost of longer program execution times. If the computation time for the duty parameters exceeds the converter switching cycle, the control loop functionality will be updated at a down-sampled rate relative to the switching rate and the sampling rate of the sensed voltage and current signals, potentially leading to signal aliasing and loop instability issues. Therefore, the control implementation target should be to keep the computation time of the optimal duty models within a switching period, which is 10 μs for the case under study. The execution time of the implemented block of code for duty cycle calculation is optimized in this work using the following methods:

1. Frequently used multiplication results are precomputed and stored in dedicated floating-point registers or memory locations. Such precomputed values are incorporated into subsequent multiplication operations to eliminate redundant recalculations and expedite execution.
2. Division operations are substituted with reciprocal multiplications whenever possible to reduce resource-intensive computation of division as it takes $5-10 \times$ more clock cycles compared to a multiplication operation.
3. Fixed-point arithmetic is evaluated if precision requirements can be met with reduced bit-width representations.

Furthermore, a trade-off analysis is performed to select a suitable order for the polynomial model of δ_k. This analysis considers both the program execution time and the compromise in switching network efficiency due to modeling errors, as presented in Table 4.6. The data suggests that a 2-D 4th order second-pass fitted model of δ_k results in the lowest average efficiency compromise of 0.12%. In comparison, 3rd and 2nd order models incur compromises of 0.24% and 0.84% in efficiency, respectively. However, benchmarking the computation time of the duty models on the TMS320F28379D DSP shows that using a 4th order model results in a program execution time of 28 μs, which is significantly higher than the switching period. Therefore, with an efficiency compromise of 0.24%, the optimal δ_k parameters are modeled using 3rd order 2-D polynomials, which only require 6.8 μs

Table 4.6 Trade-off analysis between optimal duty polynomial models with different orders

Order of the 2nd pass fitted polynomial model of δ_k	Compromise on TAB switching network efficiency (%)	Computation time by microcontroller (μs)
2	0.84	1
3	0.24	6.8
4	0.12	28

for computation. The modeling accuracy of the δ_k variables under different polynomial function orders can be observed in Fig. 4.18, where it is evident that the 2nd order δ_k models deviate more from the actual optimal δ_k values compared to the implemented 3rd order models.

Similar to the polynomial regression-based model construction for optimal duty cycles and their deployment within the control system of a DC-DC-AC TAB converter, the switching frequency (f_{sw}) can also be integrated into the control system as needed. To achieve this, an optimal f_{sw} dataset must be generated alongside the optimal δ_k values. Subsequently, a polynomial model of f_{sw} needs to be constructed. This model should function as a representation of the operating port voltages and loads, outputting the optimal f_{sw} based on the sensed parameters to minimize conduction and switching losses in the converter. Such an optimal control system construction technique is extendable to any port MAB (Multi-Active Bridge) converter, including DC-DC TABs. The usefulness of deploying polynomial models over Look-Up Tables (LUTs) or gradient descent approaches lies in their reduced memory usage. This technique becomes particularly valuable when a control system with higher degrees of freedom and a greater number of ports, with wide gain and load variation, is required. Consequently, three or higher port MABs are ideal platforms for implementing this control strategy.

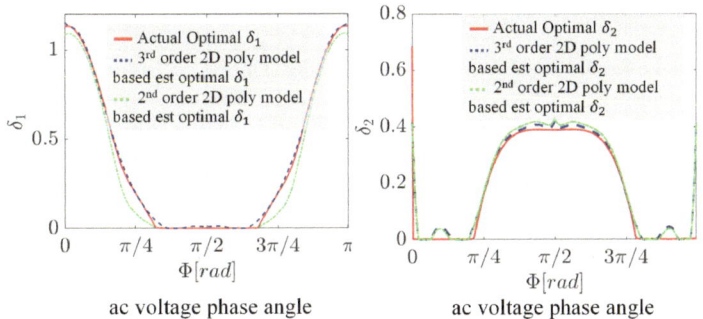

Fig. 4.18 Comparison between the optimal δ_1 and δ_2 and their models represented by different order polynomials

By employing polynomial models for both duty cycles and switching frequency, the control system can dynamically adjust to achieve optimal performance under varying conditions. This ensures minimized losses and enhanced efficiency, leveraging the computational efficiency and flexibility of polynomial regression over memory-intensive approaches.

4.5   Conclusions

This chapter provides an extensive analysis of multivariable PWM control aimed at minimizing conduction and switching losses in DC-DC-DC and DC-DC-AC TAB converters. It concludes that incorporating duty cycle control alongside phase shift control optimizes system losses by offering greater flexibility to the control system, particularly during light-load and non-unity gain operations. The application of penta-phase-shift (PPS) control, which utilizes all five degrees of freedom, enhances overall system efficiency and extends the ZVS range of the converter in wide gain power conversions. Additionally, the chapter presents a detailed mathematical framework for deriving polynomial regression models of the optimal duty cycle control parameters for a DC-DC-AC TAB converter. These models enable maximum efficiency tracking of the high-frequency TAB switching network across a broad range of AC output voltages and operating loads. Such precise models are easy to implement in digital controllers, incur minimal memory overhead, facilitate robust online tracking of the converter's loss-optimal operating point, and are extendible to other MAB converters.

References

1. A. K. Bhattacharjee, N. Kutkut, and I. Batarseh, "Review of multiport converters for solar and energy storage integration," *IEEE Trans. Power Electron.*, vol. 34, no. 2, pp. 1431–1445, Feb. 2019.
2. C. Zhao, S. D. Round and J. W. Kolar, "An Isolated Three-Port Bidirectional DC-DC Converter With Decoupled Power Flow Management," in *IEEE Transactions on Power Electronics*, vol. 23, no. 5, pp. 2443–2453, Sept. 2008, https://doi.org/10.1109/TPEL.2008.2002056.
3. S. Dey and A. Mallik, "Multivariable-Modulation-Based Conduction Loss Minimization in a Triple-Active-Bridge Converter," in *IEEE Transactions on Power Electronics*, vol. 37, no. 6, pp. 6599–6612, June 2022, https://doi.org/10.1109/TPEL.2022.3141334.
4. S. Dey, A. Mallik and A. Akturk, "Investigation of ZVS Criteria and Optimization of Switching Loss in a Triple Active Bridge Converter Using Penta-Phase-Shift Modulation," in *IEEE Journal of Emerging and Selected Topics in Power Electronics*, vol. 10, no. 6, pp. 7014–7028, Dec. 2022, https://doi.org/10.1109/JESTPE.2022.3191987.
5. M. Rashidi, N. N. Altin, S. S. Ozdemir, A. Bani-Ahmed and A. Nasiri, "Design and Development of a High-Frequency Multiport Solid-State Transformer With Decoupled Control Scheme," in *IEEE Transactions on Industry Applications*, vol. 55, no. 6, pp. 7515–7526, Nov.–Dec. 2019, https://doi.org/10.1109/TIA.2019.2939741.

6. S. Y. Kim, H. Song, and K. Nam, "Idling port isolation control of three-port bidirectional converter for EVS," *IEEE Trans. Power Electron.*, vol. 27, no. 5, pp. 2495–2506, May 2012.

7. S. Falcones, R. Ayyanar, and X. Mao, "A DC–DC multiport-converterbased solid-State transformer integrating distributed generation and storage," IEEE Trans. Power Electron., vol. 28, no. 5, pp. 2192–2203, May 2013, https://doi.org/10.1109/TPEL.2012.2215965.

8. A. A. Ibrahim, T. Caldognetto and P. Mattavelli, "Conduction Loss Reduction of Isolated Bidirectional DC-DC Triple Active Bridge," *2021 IEEE Fourth International Conference on DC Microgrids (ICDCM)*, 2021, pp. 1–8, https://doi.org/10.1109/ICDCM50975.2021.9504652.

9. P. Purgat, S. Bandyopadhyay, Z. Qin and P. Bauer, "Zero Voltage Switching Criteria of Triple Active Bridge Converter," in *IEEE Transactions on Power Electronics*, vol. 36, no. 5, pp. 5425–5439, May 2021, https://doi.org/10.1109/TPEL.2020.3027785.

10. S. Dey and A. Mallik, "An Online-Optimized ZVS-Current Tracked Soft-Switching Modulation for Triple Active Bridge Converter," in *IEEE Transactions on Power Electronics*, https://doi.org/10.1109/TPEL.2024.3429278.

11. S. Dey, A. Mallik, C. Darmody and A. Akturk, "Online Optimization of Decision Control Variables Ensuring Loss-Minima Tracking in a TAB-Based DC–AC–DC Switching Network," in *IEEE Journal of Emerging and Selected Topics in Industrial Electronics*, vol. 5, no. 3, pp. 893–907, July 2024, https://doi.org/10.1109/JESTIE.2023.3327050.

12. A. A. Ibrahim, T. Caldognetto, D. Biadene and P. Mattavelli, "Multidimensional Ripple Correlation Technique for Optimal Operation of Triple-Active-Bridge Converters," in IEEE Transactions on Industrial Electronics, vol. 70, no. 8, pp. 8032–8041, Aug. 2023, https://doi.org/10.1109/TIE.2022.3224182.

13. M. Liao, H. Li, P. Wang, T. Sen, Y. Chen and M. Chen, "Machine Learning Methods for Feedforward Power Flow Control of Multi-ActiveBridge Converters," in IEEE Transactions on Power Electronics, vol. 38, no. 2, pp. 1692–1707, Feb. 2023.

14. A. Mallik, Lecture Notes, Power Electronics EGR 494/598, Switching Loss Modeling and Calculation. https://peacelabasu.s3-us-west-1.amazonaws.com/Courses/EGR+494_598/Lecture-2+Loss+calculation.pdf.

15. Datasheet, EPC 2215, 200 V, 162 A Enhancement-Mode GaN Power Transistor, https://epc-co.com/epc/products/gan-fets-and-ics/epc2215.

16. S. Kouro, J. I. Leon, D. Vinnikov and L. G. Franquelo, "Grid-Connected Photovoltaic Systems: An Overview of Recent Research and Emerging PV Converter Technology," in *IEEE Industrial Electronics Magazine*, vol. 9, no. 1, pp. 47–61, March 2015, https://doi.org/10.1109/MIE.2014.2376976.

17. P. Morsali, S. Dey, A. Mallik and A. Akturk, "Switching Modulation Optimization for Efficiency Maximization in a Single-Stage Series Resonant DAB-based DC-AC Converter," in *IEEE Journal of Emerging and Selected Topics in Power Electronics*, https://doi.org/10.1109/JESTPE.2023.3302839.

18. T. Chen, R. Yu and A. Q. Huang, "Variable-Switching-Frequency Single-Stage Bidirectional GaN AC–DC Converter for the Grid-Tied Battery Energy Storage System," in *IEEE Transactions on Industrial Electronics*, vol. 69, no. 11, pp. 10776–10786, Nov. 2022, https://doi.org/10.1109/TIE.2021.3120483.

19. O. Kwon, K. -S. Kim and B. -H. Kwon, "Highly Efficient Single-Stage DAB Microinverter Using a Novel Modulation Strategy to Minimize Reactive Power," in *IEEE Journal of Emerging and Selected Topics in Power Electronics*, vol. 10, no. 1, pp. 544–552, Feb. 2022, https://doi.org/10.1109/JESTPE.2021.3090097.

20. S. Dey, A. Mallik and T. Warren, "A Variable Switching Frequency-Based Triple-Phase-Shifted Loss Optimized Modulation Strategy for DAB-Based DC–AC Microinverter," in *IEEE Journal*

of Emerging and Selected Topics in Power Electronics, vol. 12, no. 1, pp. 1110–1128, Feb. 2024, https://doi.org/10.1109/JESTPE.2023.3328271

21. S. Inoue and H. Akagi, "A Bidirectional DC–DC Converter for an Energy Storage System With Galvanic Isolation," in *IEEE Transactions on Power Electronics*, vol. 22, no. 6, pp. 2299–2306, Nov. 2007, https://doi.org/10.1109/TPEL.2007.909248.

22. A. K. Bhattacharjee, I. Batarseh, H. Hu and N. Kutkut, "An efficient ramp rate and state of charge control for PV-battery system capacity firming," *2017 IEEE Energy Conversion Congress and Exposition (ECCE)*, Cincinnati, OH, USA, 2017, pp. 2323–2329, https://doi.org/10.1109/ECCE.2017.8096451.

23. I. H. Hutchinson, A Student's guide to Numerical Methods, 2014.

Generalized Approach to Loss Formulation and Optimization in Multi-active Bridge Converters

5

5.1 Introduction to MAB Analytical Model

The degrees of freedom in order to control the power flow in a MAB topology are defined by the intra-bridge leg phase shift or v_i' duty ratios (δ_i), the inter-bridge phase differences (φ_i, and $\varphi_1 = 0$) or the phase differences between the different bridge voltages and the switching frequency (f_{sw}), as highlighted in the Fig. 5.2. The range of the phase-duty variables are defined as: $\delta_i \in [0, \pi/2]$; $\varphi_i \in [-\pi/2, \pi/2]$. Thus, for a n-port MAB, total number of possible control variables is $2n$, including (n − 1) phase-shifts, n duty parameters and f_{sw} (see Fig. 5.1).

In the quest to find the optimum MAB controller operating point under application of any PWM modulation technique, a multivariable multi-constrained loss optimization problem is formulated to solve for the optimum control variables. From a modeling stand-point, as explained in Chap 2, the ac bridge voltage of the ith port (reffered to port-1) can be represented as:

$$v_i(t) = \frac{4V_i}{\pi} \cdot \sum_{k=1,3,\ldots}^{\infty} \frac{1}{k} d_{ki} \sin\{k(\omega_s t - \varphi_i)\} \tag{5.1}$$

where $d_{ki} = \cos(k\delta_i)$; 'k' is the order of the harmonic, and $\omega_s = 2\pi f_{sw}$, f_{sw} being the switching frequency of the converter.

Using generalized harmonic approximation (GHA) method, the current sourced by port-i can be formulated and simplified as,

$$i_i(t) = \sum_{k=1,3,5,\ldots}^{\infty} \left(A_{i,k} \cos(k\omega_s t) + B_{i,k} \sin(k\omega_s t) \right) \tag{5.2}$$

© The Author(s), under exclusive license to Springer Nature Switzerland AG 2025 135
A. Mallik and S. Dey, *Switching Modulator Optimization in Isolated Power Converters*,
Synthesis Lectures on Power Electronics, https://doi.org/10.1007/978-3-031-81576-8_5

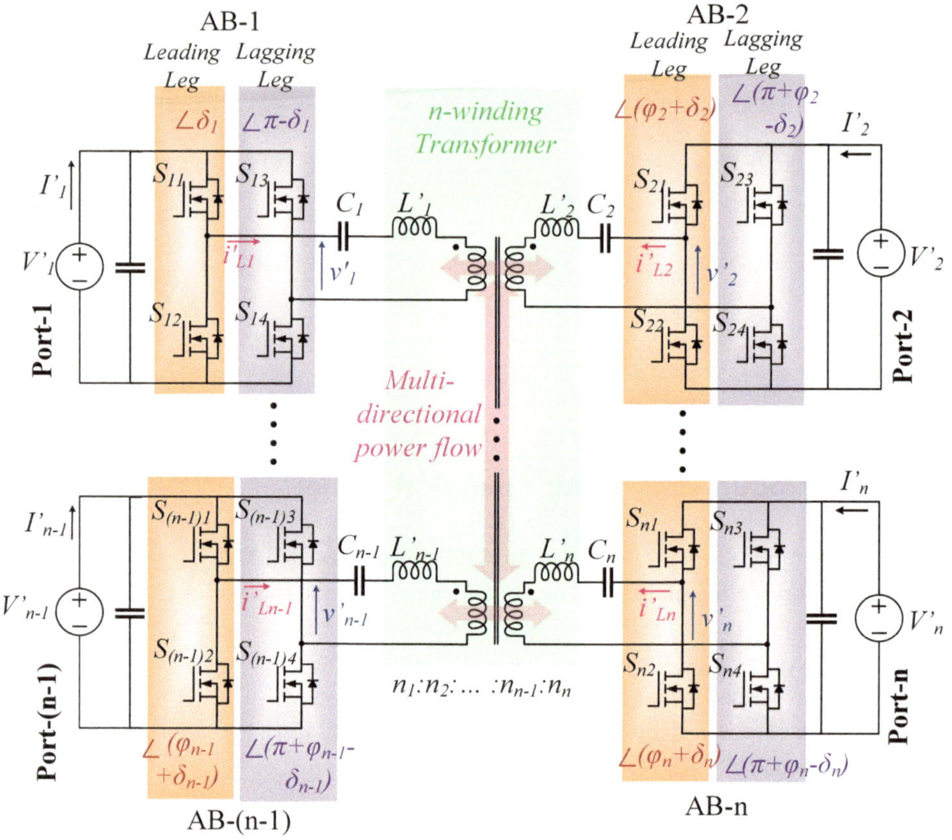

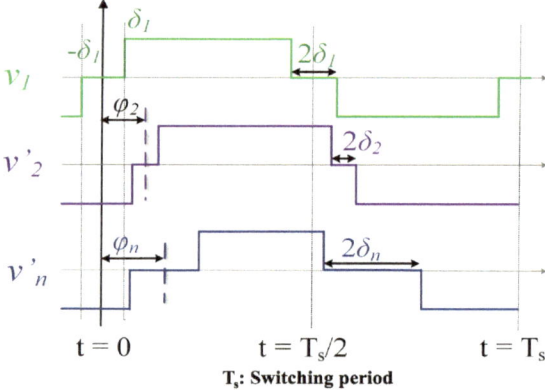

Fig. 5.1 A generic n-port MAB converter topology and phase shifts of the respective half-bridge gating signals

Fig. 5.2 Bridge voltage waveforms in MAB converter under phase-duty control

where the harmonic order dependent coefficients $A_{i,k}$ and $B_{i,k}$ are synthesized using GHA method, as detailed in Chap 2.

$$A_{i,k} = \frac{4}{\pi \omega_s k^2} \left[\sum_{\substack{j=1 \\ j \neq i}}^{n} \frac{1}{L_{ij}} \{ -V_i d_{ki} \cos(k\varphi_i) + V_j d_{kj}\cos(k\varphi_j) \} \right] \tag{5.3}$$

$$B_{i,k} = \frac{4}{\pi \omega_s k^2} \left[\sum_{\substack{j=1 \\ j \neq i}}^{n} \frac{1}{L_{ij}} \{ -V_i d_{ki} \sin(k\varphi_i) + V_j d_{kj}\sin(k\varphi_j) \} \right] \tag{5.4}$$

5.2 Formulation of the Power Loss Optimization Problem in an MAB

5.2.1 Equivalent Conduction Loss Model

The sum of squares of the rms currents flowing in the port windings (all referred to the primary side or port-1 in this case) is quantified as the generalized conduction loss objective function since it directly relates to the total conduction loss in the MAB transformer windings and switches and can be written as,

$$F_{cond}(\delta_i, \varphi_i, m_i) = \sum_{i=1}^{n} i_{i,RMS}^2 \tag{5.5}$$

where $i \in [1, n]$; $\delta_i \in [0, \pi/2]$; $\varphi_i \in [-\pi/2, \pi/2]$; $m_i = \frac{V_i}{V_1} > 0$. The winding RMS currents ($i_{i,RMS}$) can be calculated using (2.48) (refer to Chap. 2) that depends on the port voltage gain and its phase-duty control variables.

To verify the usefulness of incorporating the port voltage duty variables towards the RMS current reduction, the gradient of the defined objective function is formulated in (5.6) with respect to the port-1 duty cycle (δ_1) while keeping the rest of the ports' duty cycles as 0.

$$\nabla_{\delta_1} F_{cond} \big|_{\delta_j = 0, j \neq 1} = \frac{\partial F}{\partial \delta_1} = \sum_{i=1}^{n} \frac{\partial i_{i,RMS}^2}{\partial \delta_1} = \sum_{i=1}^{n} \left[A_{i,k=1} \frac{\partial A_{i,k=1}}{\partial \delta_1} + B_{i,k=1} \frac{\partial B_{i,k=1}}{\partial \delta_1} \right] \tag{5.6}$$

Further, to gain more insight into the mathematics while simplifying the problem, the port-1 and port-2 RMS currents' gradients are measured separately in (5.7) and (5.8) for $k = 1$. To Simplify the analysis, all the inter-port inductors of the MAB are assumed to be same as L.

$$
\nabla_{\delta_1} i_{1,RMS}^2 \big|_{\delta_j=0, j\neq 1} = \frac{\partial i_{1,RMS}^2}{\partial \delta_1} = \frac{16V_1^2}{\pi^2 \omega^2 L^2} \cdot \sin\delta_1 \cdot (n-1)
$$
$$
\cdot [-(n-1) + m_2\cos\varphi_2 + m_2\cos\varphi_2 + \cdots + m_n\cos\varphi_n] \qquad (5.7)
$$

$$
\nabla_{\delta_1} i_{2,RMS}^2 \big|_{\delta_j=0, j\neq 1} = \frac{\partial i_{2,RMS}^2}{\partial \delta_1} = \frac{16V_1^2}{\pi^2 \omega^2 L^2} \cdot \sin\delta_1
$$
$$
\cdot [1 - (n-1) \cdot m_2\cos\varphi_2 + m_3\cos\varphi_3 + \cdots + m_n\cos\varphi_n] \qquad (5.8)
$$

Now, by evaluating the gradient values at $\delta_1 = 0^+$, we can infer that for any phase shift values, $\nabla_{\delta_1} i_{1,RMS}^2 < 0$ for $m_i < 1$ for $i \in [1, n]$, whereas no conclusion regarding the port-2 RMS current gradient can be drawn under the same gain conditions. Therefore, it can be proven and suggested that the application of duty cycle variable will be able to reduce the bridge RMS currents of few ports with below-unity gains. Hence, in order to get a complete understanding of it, analyzing the duty cycle effect on the total sum square port RMS current is essential and is formulated in (5.9).

$$
\nabla_{\delta_1} F_{cond} \big|_{\delta_j=0, j\neq 1} = \frac{16V_1^2}{\pi^2 \omega^2 L^2} \cdot \sin\delta_1 \cdot n
$$
$$
\cdot [-(n-1) + m_2\cos\varphi_2 + m_2\cos\varphi_2 + \cdots + m_n\cos\varphi_n] \qquad (5.9)
$$

From (5.9), it can be concluded that the formulated conduction loss objective function has a negative gradient with respect to δ_1 for the gain condition defined by $m_i < 1$. Therefore, it is proven that with the implementation of phase-duty based higher order hybrid control, there is a scope to minimize the overall conduction loss in the MAB. This also proves the drawbacks in the only phase-shift based modulation technique, which is unable to achieve the least sum of mean square currents under a wide $\{m, P\}$ operating range.

Thus, for a specific MAB operating condition, defined by the output ports' voltage gain and load conditions (m_i, P_i), the multi-variable, multi-constrained, and single objective optimization problem can be stated as, $minF_{cond}|_{m_k}(\delta_i, \varphi_i)$ subject to a set of power transfer equality constraints.

$$
PE_i|_{m_k, P_k}(\delta_i, \varphi_i) = \sum_{\substack{j=1 \\ j\neq i}}^{n} P_{ij} - P_i = 0 \ (i = 1,2,\ldots n), \text{ where } m_k = \{m_1, m_2, m_3 \ldots m_n\}
$$
and $P_k = \{P_1, P_2, P_3 \ldots P_n\}$ for $\delta_i \in \left[0, \frac{\pi}{2}\right]$ and $\varphi_i \in \left[-\frac{\pi}{2}, \frac{\pi}{2}\right]$. Solving this problem generates the optimal control variable set $(\delta_k^*, \varphi_k^*)_{opt}$ where $\delta_k^* = \{\delta_1^*, \delta_2^*, \delta_3^* \ldots \delta_n^*\}$ and $\varphi_k^* = \{\varphi_1^*, \varphi_2^*, \varphi_3^* \ldots \varphi_n^*\}$ lead to the least conduction loss for a particular (m_k, P_k) set.

5.2.2 Equivalent Switching Loss Model and MAB ZVS Criteria

The switching losses in an MAB converter are mainly determined by the instantaneous currents through the bridge cells during their switching transitions. These switching currents can be accurately derived from the expressions of instantaneous port currents given by (5.2), which depend on the phase-shift of the gating signals for each switch. Further, it can be inferred from the MAB port voltage and current expressions presented in (5.1) and (5.2) that all the circuit waveforms are half-cycle symmetric or $v_i(t) = -v_i(\pi + t)$ and $i_i(t) = -i_i(\pi + t)$ ($i = 1, 2,$ or 3).

These relationships establish the following facts:

(a) Within any of the active bridges (ABs), switches connected in the same leg experience identical turn-on and turn-off losses separately.
(b) The turn-on current of any switch is exactly equal in magnitude to the turn-off current but flows in the opposite direction.

These connections between turn-off and turn-on switch currents are beneficial for formulating zero voltage switching (ZVS) conditions and for calculating the switching losses of MOSFETs at a specific port. Furthermore, the turn-on currents of the leading and lagging leg switches of any general port i can be expressed as $i_{Si,lead,on} = i_i(\varphi_i + \delta_i)$ and $i_{Si,lag,on} = i_i(\pi + \varphi_i - \delta_i)$, respectively. Likewise, the turn-off currents are: $i_{Si,lead,off} = -i_i(\varphi_i + \delta_i)$ and $i_{Si,lag,off} = -i_i(\pi + \varphi_i - \delta_i)$. Therefore, the equivalent turn-on and turn-off switching losses in a universal MAB can be formulated, as given in (5.10) and (5.11), correspondingly.

$$P_{i,turn\text{-}on} = |i_i(\varphi_i + \delta_i)| + |i_i(\pi + \varphi_i - \delta_i)| \tag{5.10}$$

where, $i_i(\varphi_i + \delta_i) or i_i(\pi + \varphi_i - \delta_i) < i_{i,ZVS}$.

$i_{i,ZVS}$ is the minimum instantaneous current required to facilitate ZVS and is derived in the following subsection of this chapter and also presented in detail in Chap. 2.

$$P_{i,turn\text{-}off} = |-i_i(\varphi_i + \delta_i)| + |-i_i(\pi + \varphi_i - \delta_i)| \tag{5.11}$$

The calculation of the total MAB $P_{i,turn\text{-}off}$ and $P_{i,turn\text{-}on}$ can thus be simplified using (5.2) as,

$$\sum_{i=1}^{n} P_{i,turn\text{-}off} = |-i_i(\varphi_i + \delta_i)| + |-i_i(\pi + \varphi_i - \delta_i)|$$

$$= \left| \sum_{i=1}^{n} \left[\frac{4}{\pi\omega} \sum_{\substack{j=1 \\ j \neq i}}^{n} \frac{1}{L_{ij}} \left[\sum_{k=1,3,\dots}^{\infty} \left[-\frac{V_i}{k^2} \cos^2(k\delta_i) + \frac{2V_j}{k^2} \cos(k\delta_j) \cos(k\delta_i) \cos\{k(\varphi_i - \varphi_j)\} \right] \right] \right] \right|$$

$$\tag{5.12}$$

Furthermore, when synthesizing the objective function for switching loss equivalence in a MAB network, only the peak currents during hard turn-on are considered significant, as soft turn-on (ZVS) results in near-zero switching losses at the devices. Therefore, the sum of the peak currents during hard switching is chosen as a representative metric for switching loss objective function for a given choice of power semiconductor devices in the full-bridges, as given in (5.13).

$$
\begin{aligned}
F_{sw}\left(\delta_j, \varphi_j\right) &= \sum_{i=1}^{n} P_{i,hard-turn\text{-}on} + P_{i,turn\text{-}off} \\
&= \begin{cases} \displaystyle\sum_{j=1}^{n}\left[4\left|i'_j(\varphi_j)\right| + 4\left|i'_j(\varphi_j)\right| \cdot hard(j)\right]; \delta_j = 0 \\[2ex] \displaystyle\sum_{j=1}^{n}\sum_{k=1}^{2}\left[2\left|i'_{j,k}(\tau_{j,k})\right| + 2\left|i'_{j,k}(\tau_{j,k})\right| \cdot hard(j,k)\right]; \delta_j > 0 \end{cases}
\end{aligned}
\tag{5.13}
$$

where, $hard(j, k)$ is a binary quantity that can be 0 or 1, depending on if the kth ($k = 1, 2$) half-bridge of the jth port ($j = 1, 2, 3$) undergoes ZCS or ZVS turn-on, or not; the turn-off current of kth ($k = 1, 2$) half-bridge of the jth port is shown as $\left|i_{j,k}(\tau_{j,k})\right|$, where $\tau_{j,1} = (\varphi_j - \delta_j)$ and $\tau_{j,2} = (\varphi_j + \delta_j)$. The condition for the hard-switching is primarily dependent on the port-inductor current direction as elaborated in Chap. 2.

Here, $F_{sw\,pk}(\delta_j, \varphi_j)$ is quantified in a way that the function minimization attempts to accomplish minimum sum of switching instant current values while considering all hard-switching events because the turn off losses are directly proportional to the switching instant currents.

Thus, the multi-variable and multi-constrained switching loss objective function optimization problem can be formulated as, $minF_{sw}|_{m_k}(\delta_i, \varphi_i)$ subject to $PE_i|_{m_k,P_k}(\delta_i, \varphi_i) = 0 (i = 1, 2,..., n)$. This optimization problem can be solved for the switching loss optimal control variable set $(\delta_k^*, \varphi_k^*)_{opt}$ for a converter operating condition defined by (m_k, P_k).

To accurately quantify f_{sw} as defined in (5.13), it is essential to identify the ZVS conditions in a generalized MAB converter. To achieve this, a port-equivalent MAB model is derived in this study. Figure 5.3 illustrates the step-by-step derivation of the ith port-equivalent MAB network from its Y-model, where the superposition theorem is applied to calculate the Thevenin-equivalent line inductance observed from Port-i, as described in (5.14).

$$
L_{TH,i} = \left(\sum_{\substack{j=1 \\ j \neq i}}^{n} \frac{1}{L_j}\right)^{-1} + L_i
\tag{5.14}
$$

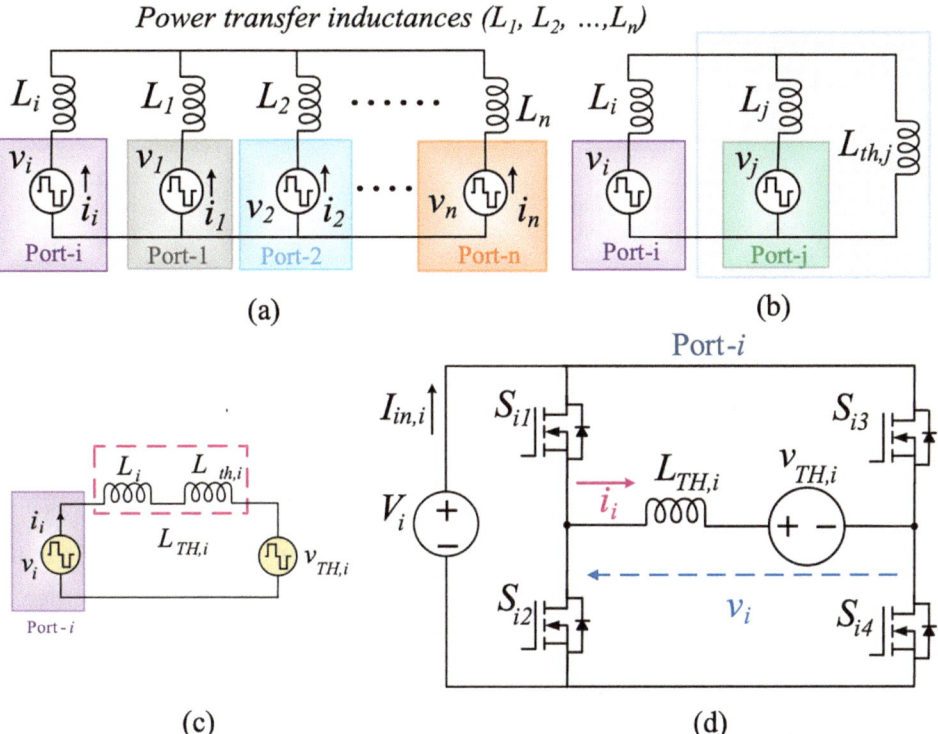

Fig. 5.3 Derivation of the Port-equivalent MAB circuit model to analyze ZVS criteria: **a** MAB Y-model; **b** Thevenin-equivalent MAB ports with respect to Port-j under superposition theorem; **c** Thevenin-equivalent MAB from perspective of Port-i; **d** Port-i equivalent MAB Circuit

The Thevenin-equivalent voltage source $v_{TH,i}$ is deduced by superposition of all the port voltage contributions and can be written as,

$$v_{TH,i} = \sum_{\substack{j=1 \\ j \neq i}}^{n} v_j \left[\frac{\left[\left(\sum_{\substack{k=1 \\ k \neq i,j}}^{n} \frac{1}{L_k} \right)^{-1} \right]}{\left[\left(\left(\sum_{\substack{k=1 \\ k \neq i,j}}^{n} \frac{1}{L_k} \right)^{-1} + L_i \right)^{-1} \right]} \right]. \tag{5.15}$$

Finally, a port-equivalent MAB structure is formulated in Fig. 5.3d. The $v_{TH,i}$ includes information regarding the phase-duty control parameters of the remaining MAB ports

Table 5.1 Summarized ZVS conditions for any port in a MAB

Case	Inclusion of δ_i	Leg undergoing turn-on event	Condition for ZVS	Switching time (τ)
1	No	Any leg	$\frac{1}{2}L_{TH,i}i_i(\tau)^2 \geq -2C_{OSS,i}V_{TH,i}(\tau)V_i$	φ_i
2	Yes	Leading leg	$\frac{1}{2}L_{TH,i}i_i(\tau)^2 \geq -2C_{OSS,i}V_{TH,i}(\tau)V_i - C_{OSS,i}V_i^2$	$\varphi_i - \delta_i$
		Lagging leg	$\frac{1}{2}L_{TH,i}i_i(\tau)^2 \geq -2C_{OSS,i}V_{TH,i}(\tau)V_i + C_{OSS,i}V_i^2$	$\varphi_i + \delta_i$

and is used to determine the ZVS conditions for the switches in the ith port AB. The derivation of the ZVS criteria for the switches in port-i is conducted based on [1] and summarized in Table 5.1. The ZVS conditions extracted from Table 5.1 are utilized in the algorithm for calculating the objective function, as previously described.

5.3   Solution of the Power Loss Optimization Problem in an MAB and Hardware Implementation of the Modulation Strategy

A flowchart depicting the implementation of the loss optimal modulator using multivariate polynomial regression models, as explained in Chaps. 3–4, is illustrated in Fig. 5.4. The phase shift variables are dedicated for regulating the port voltages or controlling the power flow to the respective ports, while the full-bridge duty ratio variables are allocated for the objective function optimization for the given set of phase shift angles (φ_k). To perform voltage regulation, (N-1) control loops are designed to produce the phase shifts as the control variables. For DC port voltage tracking, conventional proportional integral (PI) controllers can be employed, while proportional resonant (PR) or PI controllers can be used for AC sinusoid voltage tracking at any port. The φ_k variables are fed to the optimization routine that computes and minimizes the objective function, which could be conduction loss alone, or switching loss alone, or total switching network loss and determine optimized set of δ_k variables. This optimization routine can be performed offline for a set of converter design parameters, and the δ_k set can be expressed in terms of polynomial regression as a function of the port voltage gain and normalized port power levels. Assuming the converter parameters would stay unchanged during operation, the polynomial-based δ_k calculator routine can be implemented in digital control platform for performing loss minimization based on sensor feedback information.

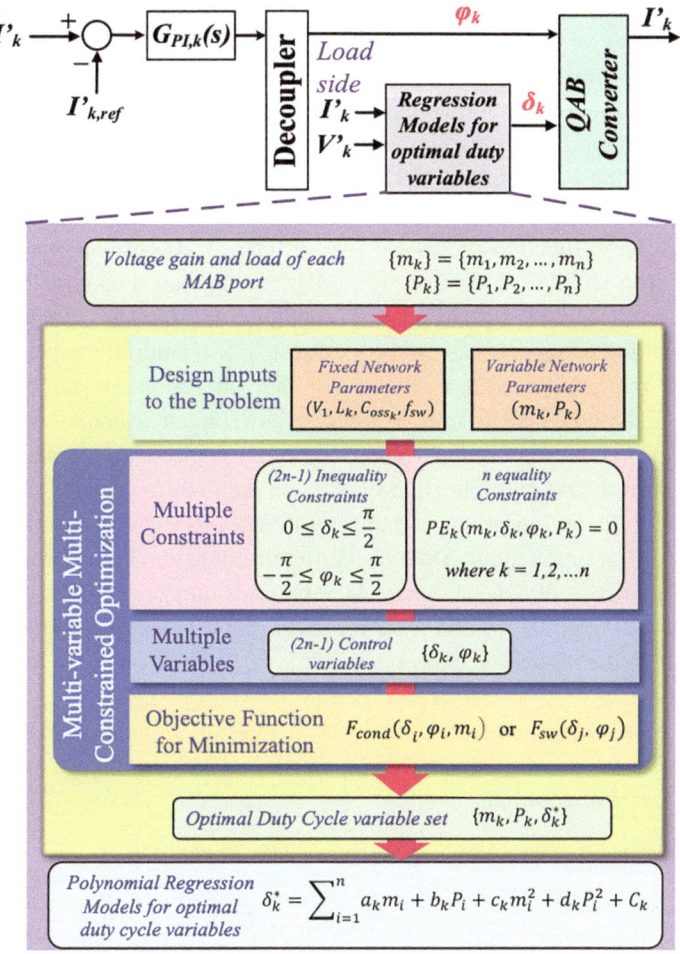

Fig. 5.4 A Flowchart depicting the implementation of the loss optimal PWM control and formulation of the regression models of the optimized duty cycle variables for a QAB

Let us take an example of a quad active bridge (QAB) dc-dc converter. In order to implement the loss-optimal modulator, the optimization routine is run to calculate the optimal control variable sets for all the combinations of the discrete gain and load values of the QAB converter ports. A ten-column matrix $(m_2, m_3, m_4, P_2, P_3, P_4, \delta_1, \delta_2, \delta_3, \delta_4)$ is formed consisting of the QAB port gain, load, and the obtained optimal duty cycle control variables. The least-mean-square error (LMSE) method is applied on this obtained dataset in order to statistically formulate the regression models of the optimal duty variables (δ_k) where δ_k is expressed as a polynomial function of the QAB ports' voltage gain and load condition: $\delta_k^* = \sum_{i=1}^{n} a_k m_i + b_k P_i + c_k m_i^2 + d_k P_i^2 + e_k P_1 m_i + C_k$, where a_k, b_k, c_k, d_k

are the system coefficients and C_k is a constant. The regression order can be chosen to be linear or quadratic or cubic or higher order, depending on the trade-off between control variable generation accuracy and computational burden to calculate them.

Computation time for such mathematical operations comprising of multiplications and additions depend on the architecture used in the controller hardware. For example, a control card containing IC TMS320F28379D follows the Reduced Instruction Set Computer (RISC)-based Instruction Set Architecture (ISA). The execution times are estimated for different regression function orders for two different C2000-series microcontrollers i.e., single-core TMS320F28335 (with a 150 MHz clock) and dual-core TMS320F28379D (with a default clock of 100 MHz and configurable clock of 200 MHz maximum). The main arithmetic operations to be performed for control variable calculation are addition and multiplication. It is found out that one addition and one multiplication for TMS320F28379D take 2 and 3 clock cycles on an average, while the same in TMS320F28335 would consume 3 and 13 cycles, respectively. Table 5.2 presents the number of addition and multiplication operations required to execute the control variable computation for the optimization routine for different types of MAB and their corresponding total execution time length. Total execution time length can be expressed as follows: $t_{exec} = M \left(N_{mult} t_{mult} + N_{add} t_{add} \right)$, where M represents number of control variables subject to optimization routine, N_{mult} and N_{add} represent the number of multiplication and addition operators in the regression expression, while t_{mult} and t_{add} denote their corresponding calculation time.

For an N-port MAB converter, a linear regression function for synthesizing each control variable would employ $2N$ terms, involving $2N$ number of multiplications and $(2N - 1)$ number of additions.

For a quadratic regression, there are $2N$ linear terms and $\left(\binom{2N}{2} + 2N \right)$ quadratic terms, hence employing $\left(2N + 2 \left(\binom{2N}{2} + 2N \right) \right)$ multiplications and $\left(2N + \left(\binom{2N}{2} + 2N \right) - 1 \right)$ additions.

For a cubic regression, there are $2N$ linear terms, $\left(\binom{2N}{2} + 2N \right)$ quadratic terms, and $\left(\binom{2N}{3} + 2N^2 + 2N \right)$ cubic terms, hence engaging $\left(2N + 2 \left(\binom{2N}{2} + 2N \right) + 3 \left(\binom{2N}{3} + 2N^2 + 2N \right) \right)$ multiplications and $\left(2N + \left(\binom{2N}{2} + 2N \right) + \left(\binom{2N}{3} + 2N^2 + 2N \right) - 1 \right)$ additions.

Table 5.2 Number of arithmetic operations for executing regression models in an $(N + 1)$-port MAB

Regression order	Linear terms	Quadratic terms	Cubic terms	Number of additions	Number of multiplications
1	$2N$	0	0	$2N$	$2N$
2	$2N$	$\left(\binom{2N}{2}+2N\right)$	0	$\left(4N+\binom{2N}{2}\right)$	$\left(2N+2\left(\binom{2N}{2}+2N\right)\right)$
3	$2N$	$\left(\binom{2N}{2}+2N\right)$	$\left(\binom{2N}{3}+2N^2+2N\right)$	$\left(6N+\binom{2N}{2}+\binom{2N}{3}+2N^2\right)$	$\left(2N+2\left(\binom{2N}{2}+2N\right)+3\left(\binom{2N}{3}+2N^2+2N\right)\right)$

Table 5.3 Execution time complexity of different MAB types for linear regression

Type of MAB	$N_{variables}$	$N_{additions}$	N_{mult}	Execution time at F28379D (100 MHz clock)	Execution time at F28335 (150 MHz clock)
DAB	3	4	4	200 ns	426 ns
TAB	5	12	12	600 ns	1.28 μs
QAB	7	24	24	1.2 μs	2.56 μs

Table 5.4 Execution time complexity of different MAB types for quadratic regression

Type of MAB	$N_{variables}$	$N_{additions}$	N_{mult}	Execution time at F28379D (100MHz clock)	Execution time at F28335 (150MHz clock) (μs)
DAB	3	10	16	680 ns	1.59
TAB	5	42	72	3 μs	7.07
QAB	7	108	192	7.9 μs	18.78

Table 5.5 Execution time complexity of different MAB types for cubic regression

Type of MAB	$N_{variables}$	$N_{additions}$	N_{mult}	Execution time at F28379D (100 MHz clock) (μs)	Execution time at F28335 (150MHz clock) (μs)
DAB	3	18	40	1.51	3.82
TAB	5	90	216	8.28	20.5
QAB	7	288	720	27.36	68.09

Execution time complexity of different MAB variants for linear, quadratic, and cubic regression types for two different microcontrollers – F28335 and F28379D are presented in Tables 5.3, 5.4, and 5.5, respectively.

The time utilized for executing these mathematical operations is the cumulative sum of the time required to implement each of these instructions. The total time $T_{execute}$ utilized in execution of these mathematical operations must be less than the switching period (T_{sw}) in order to satisfy the one cycle switching limit. If the execution time exceeds the one switching cycle limit, the control loop functionality will be updated at a downsampled rate of the sensed voltage and current that could lead to signal aliasing and loop instability issues [2]. Hence, it becomes a necessity to limit the control execution time within one switching time period. Thus, total execution time length limits the switching frequency of the converter as follows: $f_{sw,max} = 1/T_{execute}$. For example, if one wants to implement a cubic regression function in F28379D for TAB converter optimization, the switching frequency upper bound will be set at 120 kHz. On the other hand, switching frequency limit, as expected, will go further lower for a higher order converter. QAB with a cubic regression implemented in F28335 could be operated with a maximum switching frequency of 14.68 kHz. If the switching frequency happens to exceed $1/T_{execute}$, the control variables would not be updated every switching cycle, so there will be down-sampling in the main loop that might cause aliasing, signal distortion, and degraded phase margin. Therefore, for a given converter switching frequency, for a particular type of MAB converter, there is a limit to the regression order that can be realized in a given microcontroller or DSP.

Further, in order to optimize the computation time of such polynomial expressions of the duty cycles inside a DSP, some salient steps can be taken such as:

1. **Precomputation and Cache Generation**: Identify recurring multiplicative factors within the code and calculate their product values a priori during initialization or pre-processing stages. Store these results in dedicated floating-point registers or memory locations to create an efficient cache of precomputed values.
2. **Intermediate Result Utilization**: Leveraging the cached intermediate results, integrate these precomputed floating-point values into the subsequent multiplication calculations. By doing so, the microcontroller can circumvent the need for redundant multiplicative recalculations, resulting in a substantial reduction in execution time.
3. **Minimization of Division Operations**: Recognize opportunities to substitute division operations with multiplication by reciprocal values. This technique can be particularly advantageous in scenarios where division operations are resource-intensive and recurrent.
4. **Use of Fixed-Point Arithmetic**: Assess the feasibility of adopting fixed-point arithmetic instead of floating-point operations, particularly if the precision requirements can be accommodated within a reduced bit-width representation. Fixed-point arithmetic typically leads to faster execution times due to simplified hardware computations.

5.4   Power Loss Optimization Results—A QAB Case Study

The PWM control strategy for the QAB converter can contain at most twice the number of MAB ports i.e., 8 control variables that include: 3 phase-shift variables $\varphi_2, \varphi_3, \varphi_4$, 4 duty cycle variables $\delta_1, \delta_2, \delta_3, \delta_4$, and switching frequency f_{sw}. Depending on the inclusion of each duty cycle as a control parameter, the PWM modulation techniques of a QAB converter can be broadly categorized in 5 control strategies as listed here: (i) Triple Phase Shift (TPS)—$\{\varphi_2, \varphi_3, \varphi_4\}$; (ii) Quad Phase Shift (QPS)—$\{\delta_i, \varphi_2, \varphi_3, \varphi_4\}$; (iii) Penta Phase Shift (PPS)—$\{\delta_i, \delta_j, \varphi_2, \varphi_3, \varphi_4\}$; (iv) Hexa Phase Shift (HxPS) – $\{\delta_i, \delta_j, \delta_k, \varphi_2, \varphi_3, \varphi_4\}$; (v) Hepta Phase Shift (HpPS)—$\{\delta_1, \delta_2, \delta_3, \delta_4, \varphi_2, \varphi_3, \varphi_4\}$. Additionally, the switching frequency f_{sw} can be included as a variable parameter, leading to subcategories within each of these strategies: constant frequency modulation and variable frequency modulation. The mathematical framework for solving the multivariable and multi-constrained loss optimization problem in a multiport MAB DC-DC converter is shown in Fig. 5.4. This figure also illustrates the online implementation of the optimized PWM control scheme in the controller of a QAB DC-DC converter. The more duty cycle variables that are included to meet the required load demand, the more complex the PWM modulation strategy becomes, resulting in a larger set of possible control variables.

In this section, we present the loss performance of the QAB converter under different PWM modulation schemes with optimal duty parameters. By solving the multivariable loss optimization problem, we can determine the optimal control variable sets for different PWM strategies. The parameters used for the QAB circuit in this analysis are presented in Table 5.6.

Figure 5.5 illustrates the variation in the objective function value, $F_{opt,cond}$ (related to the sum of squares of the winding current RMSs), achieved by different optimal PWM strategies. This analysis is conducted under wide gain and medium load (100W) operation at port-4, with the other QAB ports operating at $\{m_2, m_3, P_2, P_3,\} = \{1, 0.8, 90W, 150W\}$. The theoretical analysis shows that at an under-unity port-4 gain of 0.8, the constant frequency $(f_{sw} = 100 \text{ kHz})$ PPS modulation significantly minimizes $F_{opt,cond}$ compared to the lower-order modulation schemes such as QPS or TPS. Similarly, Fig. 5.5b presents the minimal $F_{opt,cond}$ achieved by different QAB modulation schemes for a wide variation in m_4, with the other ports operating at $\{m_2, m_3, P_2, P_3, P_4\} = \{1, 1.2, 90, 100, 50 \text{ W}\}$.

To understand the reduction in conduction loss achieved by higher-order phase-duty modulation techniques compared to conventional phase control i.e., TPS, we quantify the

Table 5.6 Circuit parameters of QAB under study

Circuit parameters	Values
Port-1 DC link (V_1)	160 V
Port-2 DC link (V_2)	90–150, 120 V (nominal)
Port-3 DC link (V_3)	90–150, 120 V (nominal)
Port-4 DC link (V_4)	16–28, 22 V (nominal)
Port-1 related power (P_1)	400 W
Port-2 related power (P_1)	200 W
Port-3 related power (P_1)	200 W
Port-4 related power (P_1)	200 W
Transformer turns ratio (n_1: n_2: n_3: n_4)	7:5:5:1
Magnetizing inductance (L_m)	360 μH
Leakage inductances (L_1', L_2', L_3', L_4')	16, 15.4, 15.5, 0.27 μH
DC link capacitors (C_{o1}, C_{o2}, C_{o3}, C_{o4})	258, 258, 258, 398 μF
Switching frequency (f_{sw})	100 kHz

Fig. 5.5 Optimum QAB PWM technique for a wide variation in m_4, for $(m_2, m_3, P_2, P_3, P_4) = $ **a** $(1, 0.8, 90, 150, 100$ W$)$ and **b** $(1, 1.2, 90, 100, 50$ W$)$

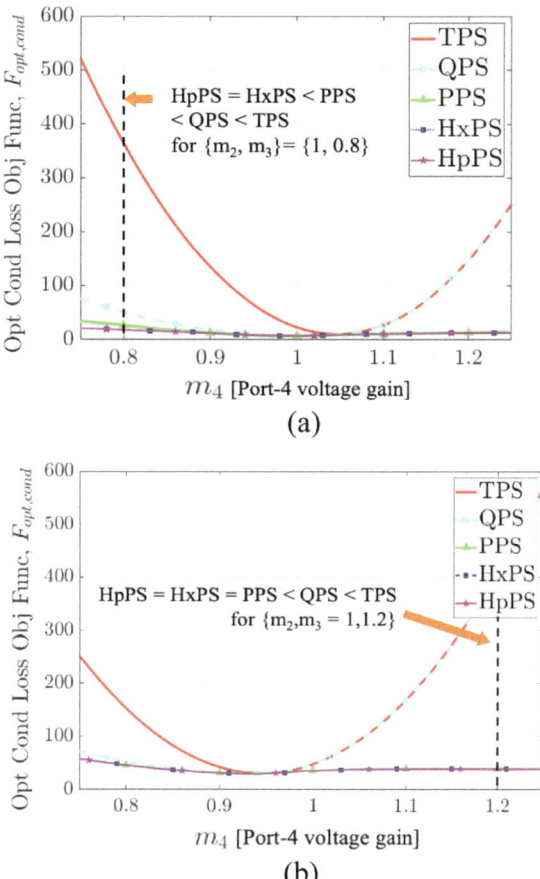

(a)

(b)

percentage reduction in optimized $F_{opt,cond}$ for different PWM schemes. The results are shown in Fig. 5.6 for varying port-4 voltage gain. The key findings are:

(i) During near unity port voltage gain operation ($m_2 = m_3 = m_4 = 1$), the conventional TPS based control can achieve the optimized conduction loss in a QAB.

(ii) As the operating point (m_4) shifts towards non-unity gain, the application of higher order control schemes becomes necessary to optimize the conduction loss.

(iii) The percentage benefit in system efficiency by employment of increasing order control system substantially rises as the load demand reduces.

(iv) The engagement of highest order control system (HpPS) in a QAB may not be very beneficial in terms of system efficiency advantage compared to the two order lower control scheme (PPS). Rather, the HpPS based control will impose more computational challenges on the controller.

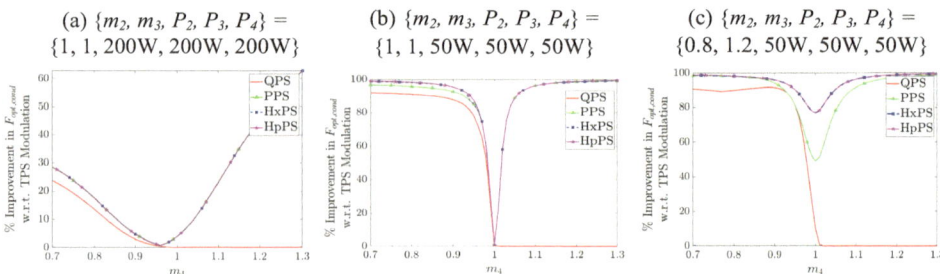

Fig. 5.6 Reduction (%) in computed conduction loss objective function ($F_{opt,cond}$) value under the application of different increasing order phase-duty control systems compared to the phase shift control alone (TPS) for varied port-4 dc link voltage gain (m_4) with three different operating conditions of rest QAB ports: **a** unity voltage gain with higher load; **b** unity voltage gain with light load; **c** non-unity voltage gain with light load

Therefore, for efficient control of a QAB, we adopt a PPS-based control scheme. This optimal PPS modulation strategy minimizes both conduction and switching losses in the QAB, achieving the best possible converter efficiency under various loading conditions. Therefore, for efficient control of a QAB, we adopt a PPS-based control scheme. This optimal PPS modulation strategy minimizes both conduction and switching losses in the QAB, achieving the best possible converter efficiency under various loading conditions.

5.5   Conclusions

This chapter presented a detailed analysis of a multi-port MAB converter circuit and a generalized model, along with a proposal for a universal optimization framework for conduction and switching loss that can be applied to any converter within the MAB family. The theoretical modeling of the MAB network is carried out using a frequency-domain GHA technique, which simplifies the calculation process by avoiding complex time-domain operations that are mode-dependent and time-consuming. Additionally, the losses in the system are accurately determined by establishing their relationships with both the instantaneous and RMS bridge currents. It is demonstrated mathematically that optimizing system losses can be achieved by implementing duty cycle control and phase shift control, which offer increased flexibility to the control system, particularly in scenarios with lighter loads and non-unity gain conditions. Furthermore, the ZVS criteria for the AB switches in a MAB are formulated based on a generic port-equivalent model, as discussed in this chapter. Moreover, in the case of an N-port MAB, a mathematical framework is developed for multivariate polynomial regression functions to synthesize control variables. This framework is then examined for its time and memory complexity to assess its feasibility for real-time microcontroller implementation. The analysis yielded specific conclusions regarding the execution time of the modulator synthesis program, and the converter switching frequency bounds.

References

1. S. Dey and A. Mallik, "Switching Network Loss Minimization Through Multivariable Modulation in a Multiactive Bridge Converter," in IEEE Transactions on Industrial Electronics, vol. 70, no. 11, pp. 10833-10847, Nov. 2023, https://doi.org/10.1109/TIE.2022.3225806.
2. Oppenheim, A.V., & Schafer, R.W. (2007). Digital Signal Processing. Prentice Hall. ISBN: 0132146355, 9780132146357.

Switching Modulator Optimization in Resonant DC-DC Converters

6

6.1 Introduction

Resonant converters are becoming more popular choices for isolated power conversion due to their inherent soft-switching capability that yield to higher efficiency and improved power density in most cases. Power density enhancement is also often attributed to the realizability of resonant inductance integration within the transformer in the form of controllable leakage. In applications with high efficiency and power density demands, resonant topologies such as LLC, CLLC, LCC, and CLL are preferred over other non-resonant DC-DC topologies due to advantages such as easier accomplishment of turn-on ZVS for source side full-bridge for almost entire power range, synchronous rectification (SR) at load side full-bridge, and unity voltage gain with sinusoidal currents at resonant frequency operation, leading to better differential mode (DM) EMI performance. Soft-switched resonant converters can be pushed towards higher switching frequency operations leading to miniaturization of magnetics while minimally impacting the converter efficiency. However, one challenge to encounter in the process is to tackle highly sensitive nature of the tank current waveforms and hence converter operation and loss performance due to any variation in circuit components or control variables. Also, the resonant converters find a wide range of applications with battery charging or photovoltaics (PV)- interfaced power conversion where the input and/or output voltage could vary in a broad range. Therefore, the converters often operate in wide voltage gain zone where the switching frequency deviates much farther away from resonant frequency, hence resulting in non-sinusoid current waveforms with higher harmonics contents. Therefore, it is critical to understand the converter power flow modeling by accounting for the unified effects of all harmonics at non-resonant frequency operation and thereby formulate the optimization problem, so that the offline modeling outcomes match the experimental behavior closely.

© The Author(s), under exclusive license to Springer Nature Switzerland AG 2025
A. Mallik and S. Dey, *Switching Modulator Optimization in Isolated Power Converters*,
Synthesis Lectures on Power Electronics, https://doi.org/10.1007/978-3-031-81576-8_6

6.2   Switching Modulator Optimization for CLLC Converter

6.2.1   Modeling and Power Flow Formulation of CLLC Converter

A resonant CLLC converter and its equivalent AC network are shown in Fig. 6.1. CLLC comprises a primary-side full-bridge in series with a resonant LC tank and a secondary-side full-bridge in series with a second LC tank isolated by a transformer with a turns ratio of N_p: N_s. Quite often, for power density driven applications, the resonant inductors are integrated in the form of controllable leakage as part of the transformer. First harmonic approximation (FHA) is an effective steady state modeling tool when the operating switching frequency is closer to the resonant frequency. However, for various applications including battery charging with wide gain requirements, the operating switching frequency is shifted further away from the resonant frequency, thus making FHA modeling infeasible for port current reconstructions. Additionally, FHA has also proven to be ineffective in zero-crossing phase angle estimation of port currents with respect to bridge voltage which is obligatory for modulating precise gate signals on secondary side active full bridge to achieve synchronous rectification (SR). Failure to achieve SR leads to heavy switching losses and hence efficiency degradation. It is also noteworthy that in some cases ZVS is conducted on the load side full-bridge to eliminate CV^2 loss unlike SR at the cost of additional turn-off losses. In recent times, several papers [1, 2] proposed modeling methods based on generalized harmonic approximation (GHA) for precise reconstruction of tank current waveforms. The AC equivalent circuits of FHA and GHA are presented in Fig. 6.2a, b. The work in [2] proposed GHA for resonant converters and experimentally proved it to be ∼ 35.5% better in port current zero-crossing phase angle estimation with respect to FHA. However, most methods approximate the AC-equivalent harmonic impedance to be resistive assuming all the harmonics of port voltage and port current to be in-phase. However, it is very well possible that even under an SR case with synchronized zero-crossings of the load-side full-bridge voltage and current, the non-fundamental voltage and current harmonics do not fall in phase due to non-sinusoid nature of the current, meaning a non-resistive harmonic impedance nature, as illustrated in Fig. 6.3. The nature of output impedance for a specific harmonic could be resistive, inductive, or capacitive depending on the polarity of voltage-to-current phase difference, albeit the zero crossings of secondary current and voltage are synchronized. Failing to account for the out-of-phase dynamics of port voltage and ports current cause inaccuracies in non-optimal control variables and hence non-optimized power losses.

   In this chapter, we introduce a revised modeling method called Improved Generalized Harmonics Approximation (I-GHA) where we propose the source/load side terminals of the inverting and rectifying full bridges as quasi-square output voltages instead of resistive impedance approximation unlike FHA or GHA and the port voltages are represented as sum of fundamental and higher order harmonics and calculated in (6.1–6.2). Additionally, for a given port, the AC-side active power injected to the rectifying/inverting full bridge

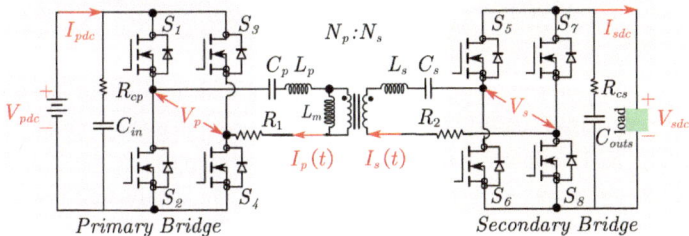

Fig. 6.1 Conventional CLLC circuit schematic

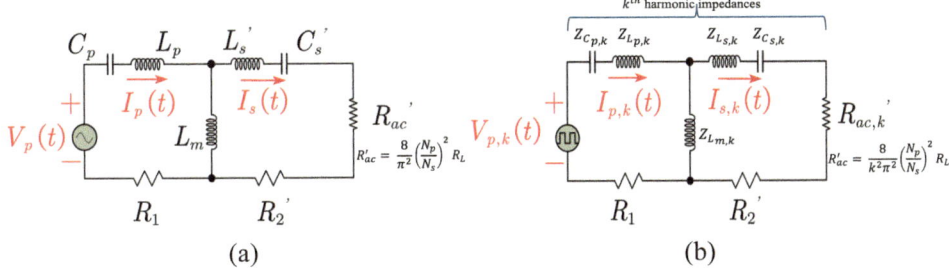

Fig. 6.2 **a** FHA based equivalent circuit, **b** GHA based equivalent circuit of CLLC

Fig. 6.3 Representation of an SR CLLC voltage-current waveforms showing non-resistive harmonic impedance effects

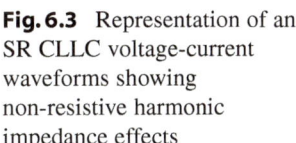

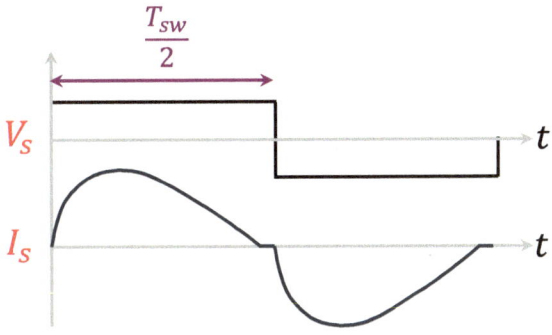

is equated to its DC output power at the output while the AC-side power is processed as the sum of the active power processed by all the individual voltage/current harmonic components, which serves as a premise for the I-GHA model.

$$V_p(t) = \frac{4V_{pdc}}{\pi} \sum_{k=1}^{2m+1} \frac{\sin(2\pi F_{sw}t)}{k} \qquad (6.1)$$

$$V_s(t)\prime = \frac{N_p}{N_s} \frac{4V_{sdc}}{\pi} \sum_{k=1}^{2m+1} \frac{\sin(2\pi F_{sw}t - k\emptyset_s)}{k} \tag{6.2}$$

The port currents and the phases corresponding to their individual harmonics are calculated by the superposition-based current summation approach shown on I-GHA equivalent network (in Fig. 6.4) adhering to the modeling principles of multi-port C3L3 in [1] adjusted for a two-port network for this application. The equivalent port impedances for kth harmonics are expressed according to (6.3). $I_{pp}(t)$ and $I_{sp}(t)$ are the portions of the current flowing through the primary LC tank when the secondary and primary ports are short-circuited, respectively. Likewise, $I_{ps}(t)$ and $I_{ss}(t)$ are the portions of the current flowing through the secondary LC tank when the secondary and primary ports are short-circuited, respectively. $Z_{p,k}$ represents the lumped kth harmonic impedance of the primary LC tank, while $Z_{s,k}$ and $Z_{m,k}$ represent the same for the secondary tank and magnetizing branch, mathematically expressed as follows, where R_1 and $R_2\prime$ represent the lumped resistance comprising of resonant capacitor ESR, transformer winding and static $R_{ds(on)}$ of semiconductor devices on primary and secondary sides respectively.

$$\left.\begin{aligned} Z_{p,k} &= j\left(k\omega L_p - \frac{1}{k\omega C_p}\right) + R_1 \\ Z_{s,k} &= j\left(k\omega L_s - \frac{1}{k\omega C_s}\right) + R_2 \\ Z_{m,k} &= Z_{L_m,k} = jk\omega L_m \end{aligned}\right\} \tag{6.3}$$

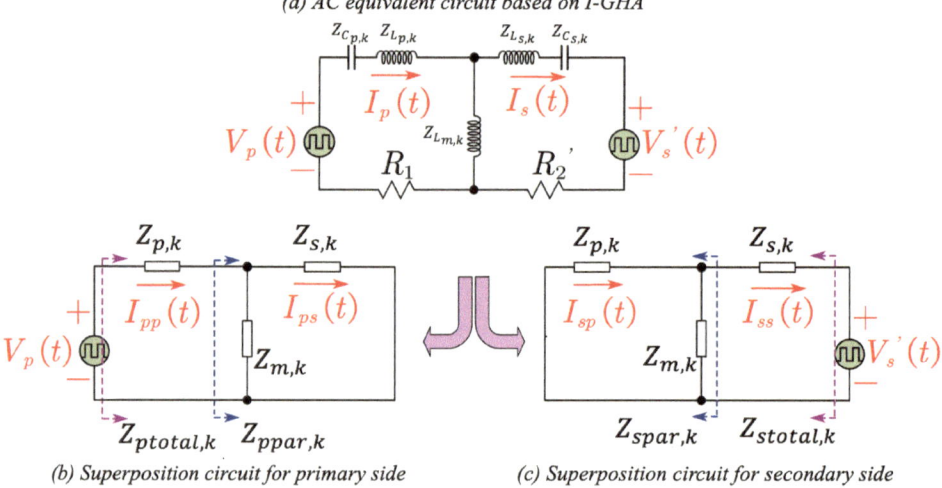

(a) AC equivalent circuit based on I-GHA

(b) Superposition circuit for primary side   (c) Superposition circuit for secondary side

Fig. 6.4 I-GHA based CLLC equivalent and superposed circuits

For the ease of analysis, the following equivalent impedances are defined in (6.4).

$$\left.\begin{aligned}
Z_{ppar,k} &= Z_{s,k} || Z_{m,k} \\
Z_{spar,k} &= Z_{p,k} || Z_{m,k} \\
Z_{ptotal,k} &= Z_{ppar,k} + Z_{p,k} \\
Z_{stotal,k} &= Z_{spar,k} + Z_{s,k}
\end{aligned}\right\} \tag{6.4}$$

After applying superposition theorem to the linear network shown in Fig. 6.4, the primary and secondary currents can be calculated as follows.

$$I_p(t) = \frac{N_p}{N_s}\left(I_{pp}(t) - I_{ps}(t)\right) \tag{6.5}$$

$$I_s(t) = \frac{N_p}{N_s}\left(I_{sp}(t) - I_{ss}(t)\right) \tag{6.6}$$

$$I_{pp}(t) = \frac{4V_{pdc}}{\pi}\sum_{k=1}^{2m+1}\frac{\sin\left(2\pi F_{sw}t - \angle Z_{ptotal,k}\right)}{k.\left|Z_{ptotal,k}\right|} \tag{6.7}$$

$$I_{ps}(t) = \frac{N_p}{N_s}\frac{4V_{sdc}}{\pi}\sum_{k=1}^{2m+1}\frac{\sin\left(2\pi F_{sw}t - k\varnothing_s + \angle Z_{spar,k} - \angle Z_{stotal,k} - \angle Z_{p,k}\right)}{k.\left|Z_{p,k}\right|}\frac{\left|Z_{spar,k}\right|}{\left|Z_{stotal,k}\right|} \tag{6.8}$$

$$I_{ss}(t) = \frac{N_p}{N_s}\frac{4V_{sdc}}{\pi}\sum_{k=1}^{2m+1}\frac{\sin\left(2\pi F_{sw}t - k\varnothing_s - \angle Z_{stotal,k}\right)}{k.\left|Z_{stotal,k}\right|} \tag{6.9}$$

$$I_{sp}(t) = \frac{4V_{pdc}}{\pi}\sum_{k=1}^{2m+1}\frac{\sin\left(2\pi F_{sw}t + \angle Z_{ppar,k} - \angle Z_{ptotal,k} - \angle Z_{s,k}\right)}{k.\left|Z_{s,k}\right|}\frac{\left|Z_{ppar,k}\right|}{\left|Z_{ptotal,k}\right|} \tag{6.10}$$

Using the reconstructed current waveforms, the active power delivered from the source side full-bridge ($P_{p,ac}$) and received at the load-side full-bridge ($P_{s,ac}$) can be calculated by adding active power contributions from all considered harmonics, as presented in (6.11–6.12). Both of those AC power expressions can be equated with the DC power sunk ($P_{s,dc}$) at the load terminal with an assumption of a lossless conversion. For a constant resistance (CR)-type load (R_L), the DC output power would be $P_{s,dc} = \frac{V_{sdc}^2}{R_L}$.

$$P_{p,ac} = \sum_{k=1}^{2m+1} V_{prms,k}I_{prms,k}\cos(-\angle I_{p,k}) \tag{6.11}$$

$$P_{s,ac} = P_{s,dc} = \sum_{k=1}^{2m+1} V_{srms,k}I_{srms,k}\cos(k\varnothing_s - \angle I_{s,k}) \tag{6.12}$$

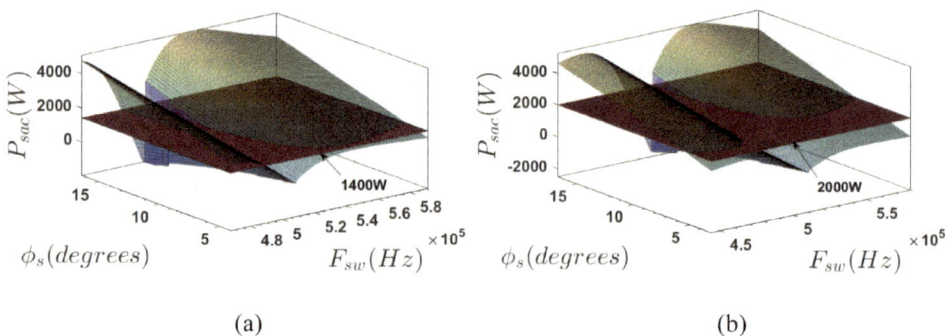

                 (a)   (b)

Fig. 6.5 P_{sac} variation with respect to control variables for $V_{sdc} = 28$ V, $V_{pdc} = 600$ V, **a** Highlighted area corresponds to 1400 W. **b** Highlighted area corresponds to 2000 W

Once $P_{p,ac}$ or $P_{s,ac}$ is equated with $P_{s,dc}$, a closed-form relation is established among voltage gain, switching frequency, and phase shift for a given load resistance. The power transfer as a function of frequency and phase shift (P_{sac} vs. F_{sw} vs. ϕ_s) is represented by 3D surfaces in Fig. 6.5 for two different cases: (a) 600–28 V at 1400 W and (b) 600–28 V at 2000 W for a 500 kHz resonant CLLC. The planes of $P_{sac} = 1400$ W and $P_{sac} = 2000$ W cut through the 3D surfaces across a phase-frequency contour, all points of which are feasible control variables for executing that voltage–power conversion. Some of the phase-frequency contour points are singled out for representative purpose and shown in Fig. 6.6. There is a visible monotonic trend between phase and frequency surrounding the resonant point i.e., 500 kHz. This can be explained by the fact that a lower frequency would likely result in a higher gain for fundamental harmonic dominance, in which case the phase shift has to be go down to decrease the fundamental power transfer and hence to maintain a specific gain. Likewise, a higher frequency would require a greater phase shift for gain regulation. This monotonicity only holds true in the vicinity of resonant operation; otherwise, the trend is no more valid due to the increasing power transfer contribution from other harmonic components.

6.2.2   I-GHA Model Verification Using a CLLC Case Study

Model verification is performed by constructing a PLECS simulation model of the circuit in Fig. 6.1a. The simulation port currents are superimposed on the analytically reconstructed port currents for a 400 V to 500 V isolated dc-dc conversion at a maximum power $P_{sdc} = 1000$ W, as shown in Fig. 6.7. Error % is calculated for analytical currents vis-à-vis the simulation currents based on two criteria: (a) RMS value and (b) switching instant value I(ζ) as they have a significant impact on the calculation of conduction

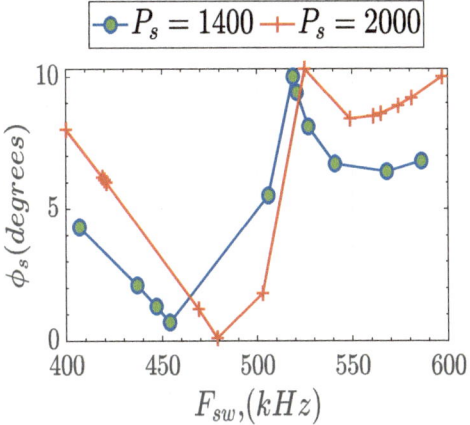

Fig. 6.6 Contours for feasible control variables representing same voltage-power conversion

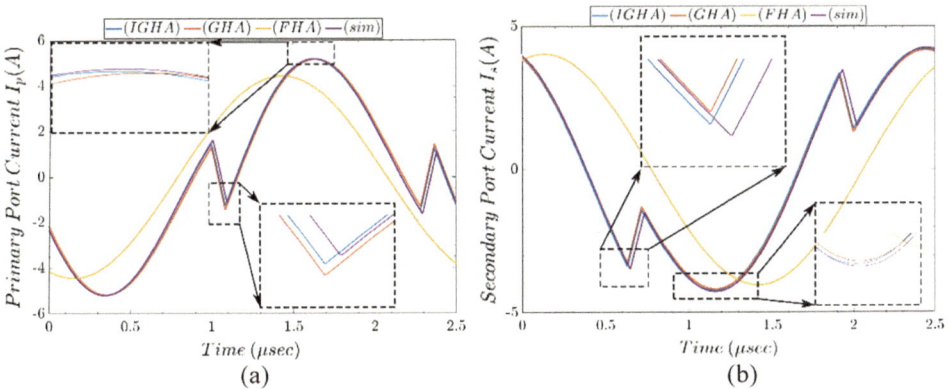

Fig. 6.7 Port currents comparison at $P_{sdc} = 1000$ W; $V_{sdc} = 500$ V; $V_{pdc} = 400$ V, $f_{sw} = 390$ kHz; $\varnothing_s = -12°$; obtained from I-GHA, GHA, FHA, and simulation for **a** I_p, **b** I_s

and switching loss, especially turn-on and turn-off loss components for the load-side full-bridge, majorly contributing to the total semiconductor loss.

Considering m = 499 for both I-GHA and GHA, Table 6.1 shows compilation of the RMS and switching instant values of the reconstructed port currents I_p, I_s for I-GHA, GHA and PLECS simulation values. FHA is not considered for comparison as the current waveshape deviates significantly from the other three outputs. I-GHA shows the lowest error % for both factors consistently showing < 5% error for both port currents. GHA could be reliably used for RMS or peak current estimation, however, for the estimation of the switching transition current (I(ζ)), the error % is considerably high for both I_p and I_s,

Table 6.1 Error % calculation for I_p, I_s @$P_{sdc} = 1000$ W; $V_{sdc} = 500$ V; $V_{pdc} = 400$ V, $f_{sw} = 390$ kHz; $\emptyset_s = -- 2°$; $N_p : N_s = 22{:}1$; $(L_p, L_s, L_m, C_p, C_s) = (7.85, 16, 73.32\ \mu H, 13.4, 6.6\ nF)$

Model	RMS (A)	RMS error %	$|I(\zeta)|$ (A)	$I(\zeta)$ error (%)
I_p(GHA)	3.63	1.3	1.65	10.4
I_p(I-GHA)	3.66	0.41	1.72	4.4
I_p(simulation—benchmark)	3.68	–	1.8	–
I_s(GHA)	2.9	2.1	3.29	5.1
I_s(I-GHA)	2.98	0.9	3.37	2.8
I_s(simulation—benchmark)	3.01	–	3.47	–

hence raising the need for employing a model like I-GHA without any AC resistive load approximation.

6.2.3 Power Loss Formulation and Optimization Using Phase-Frequency Modulation

The switching network loss includes conduction loss and turn-off losses on both full-bridges and turn-on losses that include V-I product related loss and CV^2 loss for switches with hard turn on. For ZVS FETs, only turn-off losses will be counted as part of the switching loss. The product of switching instantaneous current and frequency can be a good representative cost function of switching losses for a given choice of devices, expressed as follows for $x\varepsilon(p, s)$ [p: primary, s: secondary]: $F_{switch} = \sum_{x=p}^{s} I_x(\zeta)F_{sw}$.

By following (6.5–6.6), the primary and secondary RMS currents can be found out using I-GHA model, which then leads to the formulation of conduction loss as follows.

$$F_{cond} = \sum_{x=p}^{s} I_{rms,x}^2 R_{ds_{on},x} \qquad (6.13)$$

Hence, total semiconductor loss can be represented as:

$$F_{total} = \sum_{x=p}^{s} P_{cond,x} + P_{switch,x} + P_{qrr,x} + P_{g,x} \qquad (6.14)$$

where, $P_{qrr,x}$ denote the reverse recovery loss in the ZVS FETs of the primary and secondary side that happen due to transition of current from body diode to the FET channel as soon as the gate-to-source voltage crosses its threshold value. $P_{g,x}$ represent the gate loss that is typically the product of gate charge (Q_g), gate drive supply voltage (V_{dd}) and switching frequency.

Table 6.2 Calculation of loss functions for selected control variables at $P_{sdc} = 1000$ W, $V_{sdc} = 500$ V

Control variables (F_{sw}, $\emptyset_s$)	F_{cond}(W)	F_{switch}(A-Hz)	F_{total} (W)
(403 kHz, $-11°$)	26.73	4.79×10^6	43.1
(390 kHz, $-12°$)	25.35	4.6×10^6	10.2
(382 kHz, $-13°$)	24.85	4.49×10^6	10.62
(372 kHz, $-14°$)	24.90	4.43×10^6	11.28
(363 kHz, $-15°$)	25.66	4.44×10^6	11.99
(355 kHz, $-16°$)	26.34	4.47×10^6	12.81

$$P_{g,x} = Q_g V_{dd} F_{sw} \tag{6.15}$$

Several types of loss optimization can be performed using these cost functions depending on the designer's preference from system-level requirements. One could perform $\min(F_{cond})$ referring to minimize total conduction loss and hence minimize RMS current stress on power devices, which might allow utilization of lower current rated FETs and possible reduction in cost. One could perform $\min(F_{switch})$ or maximize the number of ZVS events for DM EMI performance enhancement. In most cases with an efficiency maximization target, the designer would likely perform $\min(F_{total})$ that would minimize the total semiconductor loss, which would also relax the thermal management needs.

As a case study example, calculation of different loss functions for 400 V to 500 V isolated dc-dc conversion at 1 kW under various feasible control variable conditions are tabulated in Table 6.2. As it is evident from the table, there are six feasible (F_{sw}, ϕ_s) control solution sets that correspond the same voltage–power combination but would result in different loss magnitudes. Thus, the final selection would be the (F_{sw}, ϕ_s) set that results in the minimum value of the cost function of designer's choice.

From RMS current minimization standpoint, (F_{sw}, ϕ_s) = (382 kHz, $-13°$) works as the most optimum solution, while for switching loss minimization, (372 kHz, $-14°$) turns out to be the best candidate and for net switching network efficiency maximization, (390 kHz, $-12°$) is found to be the most optimum control variable set. Following the similar process, cost function computation and control variable optimization can be performed in discretized operating points for all possible combinations of voltage gain and load power. The values of discrete spacings across gain and power range would determine the number of offline characterizations and optimizations to be performed on the converter using its parameter values, assuming they do not vary over time. Since the assumption of parameter invariance may not hold true practically, the parameter uncertainty estimation can be performed using artificial intelligence (AI)-driven techniques, outlined in recent literature [3] and accordingly the optimization routine needs to be updated.

6.2.4 Optimal Modulator Implementation

For a phase-frequency controlled CLLC converter without any duty modulation, the closed loop controller will generate one of the control variables, while the other will be synthesized using a LUT or regression-based approach. In most cases, switching frequency is treated as the variable generated by the controller, while the phase angle is calculated using the updated frequency and load power information provided by the voltage and current sensors. A representative control loop and its modulator implementation mechanism are illustrated in Fig. 6.8. The controller output refers to the required frequency shift that is then added to the center frequency, which is often kept the same as the resonant frequency, to derive the updated switching frequency. For a CLLC with no duty modulation, (S_1, S_4) would receive the same gate pulses that are complement of (S_2, S_3), which after the phase shift ϕ would generate (S_5, S_8) and (S_6, S_7), respectively.

For an Nth order regression, the phase shift is represented as a function of switching frequency and load power (P) in (6.16).

$$\phi = \sum_{k=0}^{N} a_k F_{sw}^k P^{N-k} \tag{6.16}$$

where the coefficients a_k are determined using least mean square error (LMSE) method applied on the offline converter characterization and optimization, as detailed in Sect. 6.3.5. The maximum executable switching frequency with this optimization framework being embedded into the DSP code can be calculated based on the total number of multiplication and addition operations and their respective time length, as elucidated in Chap. 5.

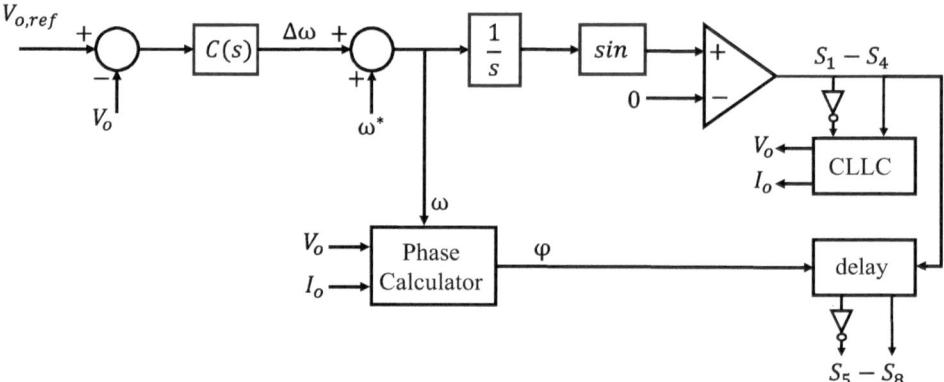

Fig. 6.8 CLLC control loop and modulator optimization mechanism [here, $V_O = V_{sdc}$ as per Fig. 6.1]

6.3   Switching Modulator Optimization for C3L3 Converter

6.3.1   Converter Operation and State-of-the-Art Modeling Techniques

The topology for a conventional bidirectional full-bridge three-port C3L3 resonant converter is presented in Fig. 6.9. It uses a multi-winding high frequency transformer (HFT), with turns (N_p, N_s, N_t) and magnetizing inductance L_m, for galvanic isolation between the source and load ports at high frequency power conversion. The primary side consists of an inverting bridge connected to the transformer primary winding through a series L-C resonant tank. The reflected AC voltages on secondary and tertiary windings are converted to DC port voltages using series LC tank-connected rectifying bridges, switching at the same frequency as the primary side. Depending on the application type, the resonant inductors (L_p, L_s, L_t) may be integrated in the controllable leakage form as part of the high-frequency transformer to enhance the converter power density and reduce the magnetic core loss. There are three resonant frequencies affiliated, one for each tank of the converter, which are expressed in (6.17).

$$\left. \begin{aligned} f_{r,p} &= \frac{1}{2\pi\sqrt{L_p C_p}} \\ f_{r,s} &= \frac{1}{2\pi\sqrt{L_s C_s}} \\ f_{r,t} &= \frac{1}{2\pi\sqrt{L_t C_t}} \end{aligned} \right\} \tag{6.17}$$

Some existing state-of-the-art modeling approaches for a three full-bridge based resonant converter topology based on Fig. 6.9 are reviewed in this section. Figure 6.10 shows the ac equivalent circuit of FHA based resonant converter circuit analysis. This modeling approach assumes the square wave outputs of primary, secondary, and tertiary full bridges as voltage sources (referred to primary) by their fundamental sinusoid equivalents by ignoring the higher order harmonics [4]. This results in a much simpler math model that eases down the digital implementation of the switching modulation. This model is fairly accurate when the converter switching frequency F_{sw} is within the vicinity of resonant frequency F_r. Operating outside a specific range beyond the resonant frequency significantly reduces the accuracy of reconstructing the port current. Incorrectly reconstructing the current in the port results in inaccurate loss calculations, which in turn leads to generating incorrect control variables, causing the converter to operate in a suboptimal condition. In converters with a lower Quality factor (Q) and narrow gain variation ($0.95 \leq G \leq 1.05$), the switching frequency (f_{sw}) is often found to be very close to the resonant frequency (f_r). In these cases, FHA has demonstrated its utility and high accuracy, making it widely adopted in both industry and academia for modeling resonant converters.

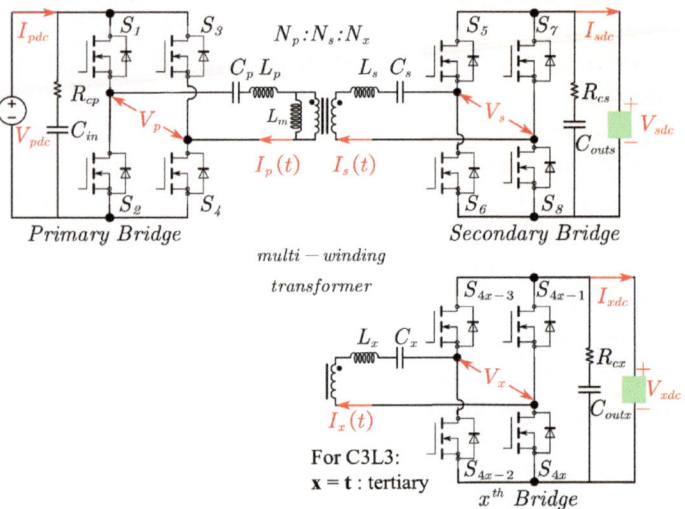

Fig. 6.9 Three-port C3L3 resonant dc-dc converter schematic

Fig. 6.10 FHA based AC equivalent circuit of C3L3

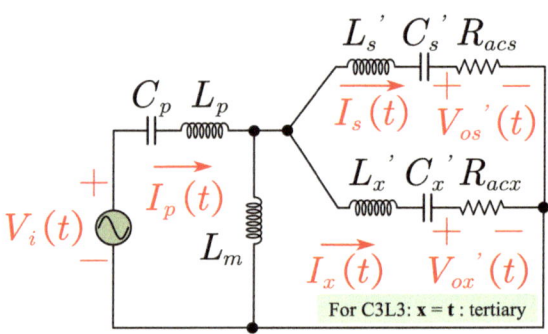

As power electronic converter applications have advanced, there has been a growing demand for DC-DC converters capable of a wider voltage gain range, especially for applications such as EV battery charging. Consequently, there has been a need for improved modeling approaches to replace FHA in port current reconstruction. Recent studies, such as those in [1, 2], have focused on developing the Generalized Harmonic Approximation (GHA). GHA incorporates higher-order switching frequency harmonics in its modeling, resulting in enhanced accuracy in port current reconstruction. However, like FHA, GHA calculates the output AC equivalent resistance by summing ratios of port voltage to port current for 'k' harmonics yet with an assumption that the voltage and current for each harmonic component would be in phase. Therefore, the terminal AC-equivalent impedance is still treated as a resistor like FHA but with superposition of a number of harmonics.

However, in reality, the harmonic components of voltage and current may not be in phase, which might make each harmonic impedance resistive or inductive or capacitive based on phase angle polarity. This error correction is incorporated in I-GHA, where the AC equivalent ports are modeled as quasi-square wave voltage sources which can be represented as a summation of **k** odd order sinusoidal voltage sources ($k \rightarrow 1$ to $2m + 1$; $m \in [1, \infty)$).

6.3.2 Power Flow Modeling Using I-GHA Technique

The major difference in I-GHA is the exclusion of the output load impedance equivalent-based model and inclusion of square wave ac voltage sources for each of the output ports based on (6.18) while it is argued that the required load impedance magnitude and phase information are contained within the port voltage formulation in the form of the control variables such as [7]: F_{sw}, ϕ_s, ϕ_t.

$$
\left.\begin{aligned}
V_p(t) &= \frac{4V_{pdc}}{\pi} \sum_{k=1}^{2m+1} \frac{\sin(2\pi F_{sw}t)}{k} \\
V_s'(t) &= \frac{N_p}{N_s} \frac{4V_{sdc}}{\pi} \sum_{k=1}^{2m+1} \frac{\sin(2\pi F_{sw}t - k\phi_s)}{k} \\
V_t'(t) &= \frac{N_p}{N_t} \frac{4V_{tdc}}{\pi} \sum_{k=1}^{2m+1} \frac{\sin(2\pi F_{sw}t - k\phi_t)}{k}
\end{aligned}\right\}
\tag{6.18}
$$

Here $V_s'(t)$ and $V_t'(t)$ represent the reflected secondary and tertiary port square-wave voltages on the primary side, respectively. In ideal multi-port converters, power can freely flow between ports bidirectionally. This means that the sum of DC output powers always equals the sum of DC input powers for a lossless power conversion. Furthermore, for any given port, the AC active power injected into the rectifying or inverting full bridge can be equated to its corresponding DC port power. In the process of formulating the tank current expressions, the resistive non-idealities of the semiconductor devices, capacitors, and magnetic elements are taken into account. These fundamental assertions lay the groundwork for developing the I-GHA model, whose AC equivalent representation is depicted in Fig. 6.11a. The objective of the forthcoming derivation is to establish the time-domain dynamics of port currents and subsequently formulate the expressions for AC active port power. Say, P_1 is the power sourced from port-1, and P_2 and P_3 denote the power sunk at the ports-2, 3. $P_{ij} = -P_{ji}$ represents the active power transferred from port i to port j. Therefore, the following can be written: $P_1 = P_{12} + P_{13}$; $P_2 = P_{12} + P_{32}$; $P_3 = P_{23} + P_{13}$.

As observed in Fig. 6.11, due to multiple voltage sources in a linear circuit, Superposition based port current formulation approach is utilized for obtaining port current equations following the steps below. Port currents defined in Fig. 6.11a are calculated as cumulative sum of individual currents in'x' loops as shown in Fig. 6.11b–d as follows:

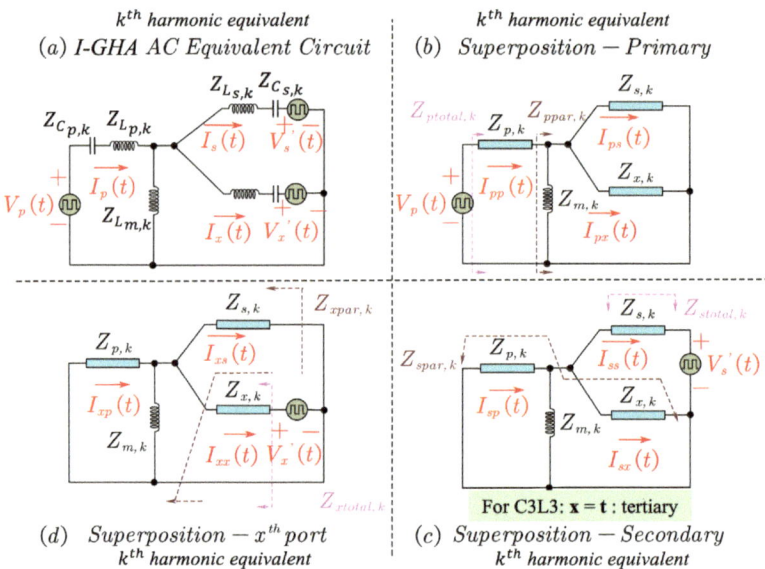

Fig. 6.11 **a** I-GHA ac equivalent model, **b–d** equivalent port circuits post applying Superposition at primary, secondary, and tertiary ports, respectively

$$
\left.
\begin{aligned}
I_p(t) &= \left(I_{pp}(t) + I_{ps}(t) + I_{pt}(t)\right) \\
I_s(t) &= \frac{N_p}{N_s}\left(I_{ss}(t) + I_{sp}(t) + I_{st}(t)\right) \\
I_t(t) &= \frac{N_p}{N_t}\left(I_{tt}(t) + I_{ts}(t) + I_{tp}(t)\right)
\end{aligned}
\right\}
\tag{6.19}
$$

In all three sides of the switching network, the resistive non-idealities are captured within lumped equivalents i.e., R_p, R_s, and R_t that represent the summation of on-resistances ($R_{ds,on,x}$) of conducting MOSFETs, resonant capacitor ESR (ESR_x), and AC + DC resistance ($R_{winding,x}$) of the transformer winding on primary (p), secondary (s), and tertiary (t) sides, respectively, for $x \in [p, s, t]$. According to the diagram, $R_p = 2R_{ds,on,p} + R_{winding,p} + ESR_p$; $R_s = 2R_{ds,on,s} + R_{winding,s} + ESR_s$; $R_t = 2R_{ds,on,t} + R_{winding,t} + ESR_t$.

The resonant port impedance Z_x for $x \in [p, s, t]$ and magnetizing impedance Z_m for kth harmonic are calculated in (6.20–6.22):

$$
Z_{p,k} = R_p + j2\pi F_{sw}kL_p - j\frac{1}{2\pi F_{sw}kC_p}
\tag{6.20}
$$

$$
Z_{s,k} = \left(\frac{N_p}{N_s}\right)^2 \left(R_s + j2\pi F_{sw}kL_s - j\frac{1}{2\pi F_{sw}kC_s}\right)
\tag{6.21}
$$

$$Z_{t,k} = \left(\frac{N_p}{N_t}\right)^2 \left(R_t + j2\pi F_{sw}kL_t - j\frac{1}{2\pi F_{sw}kC_t}\right) \tag{6.22}$$

For the sake of calculation simplicity, several impedance functions (of kth harmonic) are defined here:

$$Z_{ppar,k} = Z_{s,k}||Z_{t,k}||Z_{m,k} \tag{6.23}$$

$$Z_{spar,k} = Z_{p,k}||Z_{t,k}||Z_{m,k} \tag{6.24}$$

$$Z_{tpar,k} = Z_{p,k}||Z_{s,k}||Z_{m,k} \tag{6.25}$$

$$Z_{p,total,k} = Z_{p,k} + Z_{ppar,k} \tag{6.26}$$

$$Z_{s,total,k} = Z_{s,k} + Z_{spar,k} \tag{6.27}$$

$$Z_{t,total,k} = Z_{t,k} + Z_{tpar,k} \tag{6.28}$$

The branch currents in Fig. 6.10b are constructed in (6.29)–(6.31):

$$I_{pp}(t) = V_{pdc} \sum_{k=1}^{2m+1} \frac{\sin(2\pi F_{sw}kt - \angle Z_{ptotal,k})}{k|Z_{ptotal,k}|} \tag{6.29}$$

$$I_{ps}(t) = V'_{sdc} \sum_{k=1}^{2m+1} \frac{|Z_{spar,k}|}{|Z_{stotal,k}|} \frac{\sin(2\pi F_{sw}kt - k\phi_s + \angle Z_{spar,k} - \angle Z_{p,k} - \angle Z_{stotal,k})}{k|Z_{p,k}|} \tag{6.30}$$

$$I_{pt}(t) = V'_{tdc} \sum_{k=1}^{2m+1} \frac{|Z_{tpar,k}|}{|Z_{ttotal,k}|} \frac{\sin(2\pi F_{sw}kt - k\phi_t + \angle Z_{tpar,k} - \angle Z_{p,k} - \angle Z_{ttotal,k})}{k|Z_{p,k}|} \tag{6.31}$$

The branch currents in Fig. 6.10c are constructed as follows:

$$I_{sp}(t) = V_{pdc} \sum_{k=1}^{2m+1} \frac{|Z_{ppar,k}|}{|Z_{ptotal,k}|} \frac{\sin(2\pi F_{sw}kt + \angle Z_{ppar,k} - \angle Z_{s,k} - \angle Z_{ptotal,k})}{k|Z_{s,k}|} \tag{6.32}$$

$$I_{ss}(t) = V'_{sdc} \sum_{k=1}^{2m+1} \frac{\sin(2\pi F_{sw}kt - k\phi_s - \angle Z_{stotal,k})}{k|Z_{stotal,k}|} \tag{6.33}$$

$$I_{st}(t) = V'_{tdc} \sum_{k=1}^{2m+1} \frac{|Z_{tpar,k}|}{|Z_{ttotal,k}|} \frac{\sin(2\pi F_{sw}kt - k\phi_t + \angle Z_{tpar,k} - \angle Z_{s,k} - \angle Z_{ttotal,k})}{k|Z_{s,k}|} \tag{6.34}$$

The branch currents in Fig. 6.10d are constructed as follows:

$$I_{tp}(t) = V_{pdc} \sum_{k=1}^{2m+1} \frac{|Z_{ppar,k}|}{|Z_{ptotal,k}|} \frac{\sin(2\pi F_{sw}kt + \angle Z_{ppar,k} - \angle Z_{t,k} - \angle Z_{ptotal,k})}{k|Z_{s,k}|} \tag{6.35}$$

$$I_{ts}(t) = V'_{sdc} \sum_{k=1}^{2m+1} \frac{|Z_{spar,k}|}{|Z_{stotal,k}|} \frac{\sin(2\pi F_{sw}kt - k\phi_s + \angle Z_{spar,k} - \angle Z_{t,k} - \angle Z_{stotal,k})}{k|Z_{t,k}|} \tag{6.36}$$

$$I_{tt}(t) = V'_{tdc} \sum_{k=1}^{2m+1} \frac{\sin(2\pi F_{sw}kt - k\phi_t - \angle Z_{ttotal,k})}{k|Z_{ttotal,k}|} \tag{6.37}$$

Mean square current for any tank is calculated by adding the square of amplitudes of all different harmonic components in the reconstructed waveform by following (6.29)–(6.37) and dividing the result by 2. As an example, for primary side the net RMS current value is expressed as follows.

$$I_{p,RMS} = \frac{1}{2} \sqrt{\begin{bmatrix} \left(V_{pdc}\sum_{k=1}^{2m+1}\frac{1}{k|Z_{ptotal,k}|}\right)^2 + \left(V'_{sdc}\sum_{k=1}^{2m+1}\frac{|Z_{spar,k}|}{|Z_{stotal,k}|}\frac{1}{k|Z_{p,k}|}\right)^2 + \left(V'_{tdc}\sum_{k=1}^{2m+1}\frac{|Z_{tpar,k}|}{|Z_{ttotal,k}|}\frac{1}{k|Z_{p,k}|}\right)^2 \\ +V_{pdc}V_{sdc'}\sum_{k=1}^{2m+1}\frac{|Z_{spar,k}|}{k^2|Z_{p,k}||Z_{ptotal,k}||Z_{stotal,k}|}\cos(\angle Z_{ptotal,k}-k\phi_s+\angle Z_{spar,k}-\angle Z_{p,k}-\angle Z_{stotal,k}) \\ +V_{pdc}V_{tdc'}\sum_{k=1}^{2m+1}\frac{|Z_{tpar,k}|}{k^2|Z_{p,k}||Z_{ptotal,k}||Z_{ttotal,k}|}\cos(\angle Z_{ptotal,k}-k\phi_t+\angle Z_{tpar,k}-\angle Z_{p,k}-\angle Z_{ttotal,k}) \\ +V_{sdc}V_{tdc'}\sum_{k=1}^{2m+1}\frac{|Z_{tpar,k}||Z_{spar,k}|}{k^2|Z_{p,k}|^2|Z_{stotal,k}||Z_{ttotal,k}|}\cos(k\phi_s-k\phi_t+\angle Z_{tpar,k}-\angle Z_{spar,k}+\angle Z_{stotal,k}-\angle Z_{ttotal,k}) \end{bmatrix}} \tag{6.38}$$

Total active AC port power injected to the transformer from the primary side inverting full-bridge can be expressed as a sum of active power contributions from all harmonic components considered, as follows.

$$P_{pac} = \sum_{k=1}^{2m+1} V_{prms,k} I_{prms,k} \cos(\angle I_{p,k}) \tag{6.39}$$

Likewise, the AC active power fed to the secondary and tertiary rectifying bridges are expressed by (6.40–6.41).

$$P_{sac} = \sum_{k=1}^{2m+1} V_{srms,k} I_{srms,k} \cos(k\phi_s - \angle I_{s,k}) \tag{6.40}$$

$$P_{tac} = \sum_{k=1}^{2m+1} V_{trms,k} I_{trms,k} \cos(k\phi_t - \angle I_{t,k}) \tag{6.41}$$

The summation of active power at all output ports is equal to the input active power assuming the converter having one source and two sink ports and the conversion to be lossless, i.e., $P_{pac} = P_{sac} + P_{tac}$.

Individually, P_{sac} and P_{tac} can be equated with their respective DC port power that are fed to the loads, expressed as follows for a resistive load case, where R_S and R_T denote the load resistors at the secondary and tertiary ports, respectively.

$$P_{sac} = \sum_{k=1}^{2m+1} V_{srms,k} I_{srms,k} \cos(k\phi_s - \angle I_{s,k}) = \frac{V_{sdc}^2}{R_S} \tag{6.42}$$

$$P_{tac} = \sum_{k=1}^{2m+1} V_{trms,k} I_{trms,k} \cos(k\phi_t - \angle I_{t,k}) = \frac{V_{tdc}^2}{R_T} \tag{6.43}$$

For constant current (CC) or constant voltage (CV) types of load, the DC power at a port can simply be calculated using voltage-current product. From (6.42–6.43), a clear correlation can be established among the voltage gains, load power, and the control variables i.e., switching frequency and phase shifts.

6.3.3 I-GHA Model Verification and Its Comparison with FHA and GHA Counterparts

A three-port conversion rated for 2 kW is considered as a case study to validate the I-GHA model. Let us assume that the primary port voltage is 400 V and secondary port voltage (V_{sdc}) ranges from 500 to 600 V, while the tertiary port voltage (V_{tdc}) is varying from 22 to 28 V. This case study uses the following parameter specifications: (N_p:N_s:N_t) = (16:22:1), (L_p, L_s, L_t, L_m, C_p, C_s, C_t) = (7.85μH, 16μH, 68nH, 73.32 μH, 13.4 nF, 6.67 nF, 1.65 μF). The I-GHA generated port currents are examined against the conventional techniques i.e. GHA and FHA and compared to the simulation obtained currents in Figs. 6.12, 6.13 and 6.14 at pre-defined voltage–power levels. Figure 6.12 represents 400 V-600 V-28 V conversion at ($P_s = 1$ kW; $P_T = 1$ kW). Figure 6.13 illustrates 400–500–22 V conversion at ($P_s = 1$ kW; $P_T = 1$ kW), while Fig. 6.14 represents 400–600–22 V conversion at ($P_s = 100$ W; $P_T = 100$ W). Table 6.3 summarizes the maximum values of port currents at switching instants, along with the conduction and switching losses of the switching network. The percentage error magnitudes (in parentheses) for conduction and switching losses are also computed based on simulations performed in PLECS software. Across nearly all cases, the comparison between simulated and analytically formulated port currents reveals that I-GHA demonstrates significantly lower error percentages compared to GHA. FHA is excluded from this comparison due to considerable waveform deviations from simulated currents. Particularly notable is the higher discrepancy between switching instance values of port currents estimated by I-GHA and GHA, especially for the tertiary port. This is attributed to lower tertiary side port AC impedance resulting from a reduced turns ratio, which causes substantial variations in the estimation of switching instant currents. Despite GHA being acceptable for estimating peak and RMS currents with a mean error percentage of approximately 5%, it leads to considerable deviations in switching instantaneous current estimation. Moreover, in terms of loss calculation, I-GHA shows significantly lower error percentages for both conduction and switching losses compared to GHA.

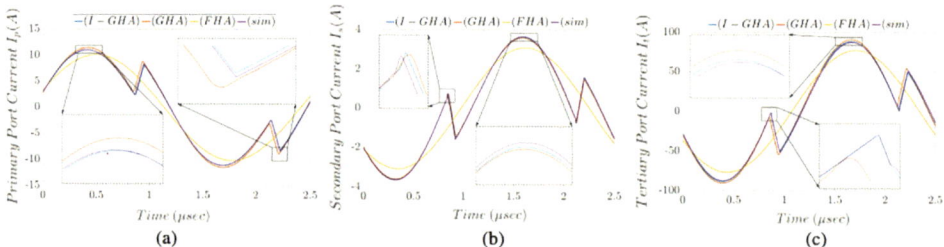

Fig. 6.12 Comparison of analytically reconstructed port currents {Ip, Is, Ix} from I-GHA, GHA and FHA with simulation port currents at {600, 28 V, 1000, 1000 W} with {398k Hz, − 25.074°, − 23.49°}

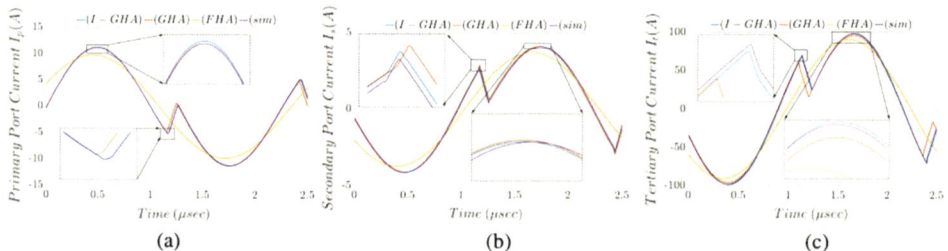

Fig. 6.13 Comparison of analytically reconstructed port currents {Ip, Is, Ix} from I-GHA, GHA and FHA with simulation port currents at {500, 22 V, 1000, 1000 W} with {396 kHz, − 12.6°, − 14°}

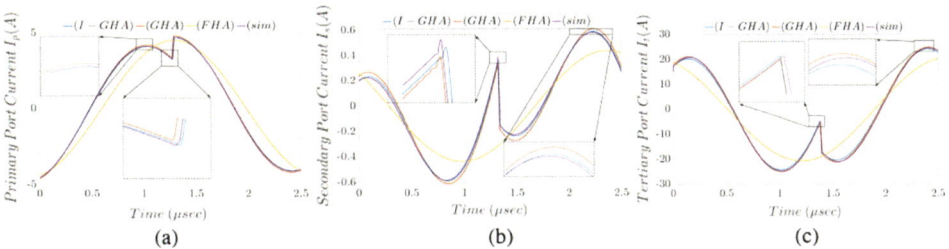

Fig. 6.14 Comparison of analytically reconstructed port currents {Ip, Is, Ix} from I-GHA, GHA and FHA with simulation port currents at {600, 22 V, 100, 100W} with {360 kHz, − 2.6°, − 2.2°}

6.3.4 I-GHA-Derived Semiconductor Loss Model Construction and Optimization

A. Loss model construction

Using the RMS current expression established in (6.38) and similar equivalents for other tank currents, the conduction losses in all three sides of the converter can be formulated. For switching loss calculation, soft-switching constraints need to be first established and then accordingly turn-on losses, as applicable, will be accounted. The three main constraints for achieving ZVS in a three-port C3L3 resonant converter are (a) lagging port current phase with respect to port voltage, (b) adequate minimum dead time in a half-bridge leg, and (c) a minimum threshold of magnetizing current. Realizability of the first constraint is dependent on the operational control variable set and is satisfied by operating the converter in the negative slope of gain vs frequency curve, also referred to as the inductive region [5]. Second and third constraints are inter-related. An equivalent three-port model is developed to analytically investigate all semiconductor devices at various switching instants within a switching cycle. Similar analysis has been formulated for a two-port full-bridge CLLC near resonance to obtain the relationship of dead time t_d, magnetizing inductance value L_m, switching frequency F_{sw} and charge-equivalent device output capacitance $C_{oss,q}$ as [6]: $L_m = \frac{t_d}{8F_{sw}C_{oss,q}}$.

The switch nomenclature used in Fig. 6.15 corresponds to primary, secondary and tertiary-side power devices as follows: $S_J \in \{S_1, S_5, S_9\}$; $S_L \in \{S_3, S_7, S_{11}\}$; $S_K \in \{S_2, S_6, S_{10}\}$; $S_M \in \{S_4, S_8, S_{12}\}$.

Here we perform the ZVS analysis by transforming all elements to the primary side. The Thevenin equivalent parameters i.e., Thevenin equivalent voltage ($V_{p,TH}$) and Thevenin impedance ($Z_{p,TH}$), according to Fig. 6.15 can be calculated for primary port as follows and can be extended to other ports.

$$\widetilde{V_{pTH,k}} = \widetilde{V_{s,k}} \frac{Z_{m,k}||Z_{t,k}}{Z_{m,k}||Z_{t,k} + Z_{s,k}} + \widetilde{V_{t,k}} \frac{Z_{m,k}||Z_{s,k}}{Z_{m,k}||Z_{s,k} + Z_{t,k}} \tag{6.44}$$

Table 6.3 Comparison of reconstructed port currents based on I-GHA, GHA and simulation values for three voltage–power combinations with V_{pdc} fixed at 400 V (the numbers in parentheses represent absolute error percentages)

600, 28 V, 1, 1 kW	Simulation	I-GHA	GHA
$I_{ppk}(A)$	11.08 (0)	11.06 (0.1)	11.49 (3.7)
$I_{spk}(A)$	3.655 (0)	3.61 (1.1)	3.59 (1.7)
$I_{tpk}(A)$	87.01 (0)	87.69 (0.7)	90.17 (3.6)
$I_p(\zeta)(A)$	8.35 (0)	8.2 (1.7)	8.89 (6.5)
$I_s(\zeta)(A)$	0.71 (0)	0.74 (3.6)	0.73 (2.2)
$I_t(\zeta)(A)$	1.14 (0)	1.17 (2.1)	4.77 (104)
$P_{cond}(W)$	8.9 (0)	8.9 (0)	9.5 (6.8)
$P_{switch}(W)$	22.7 (0)	22.6 (0.5)	24.7 (8.9)
500, 22 V, 1, 1 kW	Simulation	I-GHA	GHA
$I_{ppk}(A)$	11.2 (0)	11.26 (0.5)	11.26 (0.5)
$I_{spk}(A)$	4.1 (0)	4.10 (0.4)	4.12 (0.5)
$I_{tpk}(A)$	97.96 (0)	98.37 (0.4)	95.84 (2.1)
$I_p(\zeta)(A)$	5.15 (0)	5.2 (1)	4.8 (6.7)
$I_s(\zeta)(A)$	2.7 (0)	2.81 (4)	2.89 (4)
$I_t(\zeta)(A)$	70.86 (0)	68.96 (2.6)	61.16 (13.6)
$P_{cond}(W)$	9.8 (0)	9.8 (0.2)	9.6 (1.5)
$P_{switch}(W)$	27.1 (0)	27.1 (0.3)	25.5 (5.9)
600, 22 V, 100 W, 100W	Simulation	I-GHA	GHA
$I_{ppk}(A)$	4.12 (0)	4.1 (0.4)	4.21 (2.1)
$I_{spk}(A)$	0.59 (0)	0.58 (0.8)	0.61 (3.7)
$I_{tpk}(A)$	24.65 (0)	24.21 (1.7)	24.95 (1.2)
$I_p(\zeta)(A)$	3.28 (0)	3.272 (0.03)	3.36 (2.5)
$I_s(\zeta)(A)$	0.39 (0)	0.37 (5.1)	0.34 (12.8)
$I_t(\zeta)(A)$	4.72 (0)	4.71 (0.2)	5.37 (13.7)
$P_{cond}(W)$	1.0 (0)	1.01 (1)	1.07 (7)
$P_{switch}(W)$	15 (0)	15.6 (3.8)	15.7 (4.67)

$$Z_{p,TH,k} = Z_{p,k} + Z_{m,k} \| Z_{s,k} \| Z_{t,k} \qquad (6.45)$$

The Thevenin port voltage (referred to the primary) time domain expression can be established as follows.

Fig. 6.15 ZVS investigation using Thevenin equivalent circuit **a** for switches (S$_L$, S$_K$), current direction shown in green and V$_{yeq}$(t) > 0, I$_y$(t) > 0 **b** for switches (S$_J$, S$_M$), current direction shown in blue and V$_{yeq}$(t) < 0, I$_y$(t) < 0

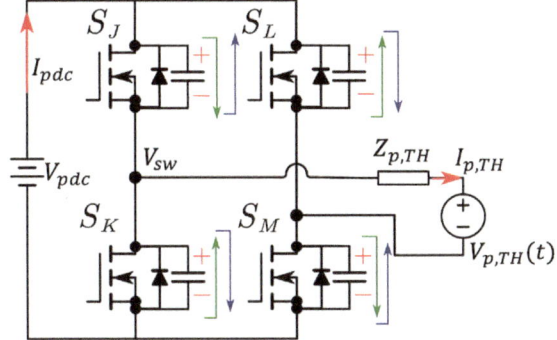

$$V_{pTH}(t) = \sum_{k=1}^{\infty} \left| \widetilde{V_{pTH,k}} \right| \sin\left(2\pi fkt + \measuredangle\widetilde{V_{pTH,k}}\right) \tag{6.46}$$

ZVS criteria for two distinct commutation cases for both the inverting and rectifying modes for source-side and load-side bridge, respectively, are explained as follows.

For evaluating inverting-side soft-switching, we follow the instantaneous circuit structure shown in Fig. 6.15 and use the turn-on instant currents for switches S_L and S_K at switching instant (t = ζ). Thereby, the total energy sunk by the Thevenin port voltage $V_{p,TH}$ can be expressed as:

$$E_{sunk} = \int_{t^*}^{t^*+t_{ZVS}} \left[V_{p,TH}(t)I_{p,TH}(t) - V_{pdc}(t)I_{pdc}(t)\right]dt$$

$$= \int_{0}^{t_{ZVS}} \left[V_{p,TH}\left(2C_{OSS,p}\frac{dV_{sw}}{dt}\right) - 0\right]dt$$

$$= 2C_{OSS,p}V_{p,TH}(t^*)V_{pdc} \tag{6.47}$$

where t_{ZVS} is the time required to completely discharge the charge-equivalent FET body capacitances $C_{OSS,p}$ of the primary side devices. To facilitate ZVS, the difference in total energy stored in the reactive elements across the switching transition as part of the primary-referred Thevenin impedance must be greater than the energy sunk by the Thevenin port voltage $V_{p,TH}$. That formulates the necessary constraint: $E_{sourced} \geq E_{sunk}$ for facilitating ZVS per (6.48) where $V_{Z_{TH}}$ is the voltage appearing across the Thevenin impedance. This relation would help us determine the minimum current $I_{ZVS,min} = |I_{p,TH}|$ required for ZVS activation on switches S_L and S_K. Similarly, since the bridge voltages are half-wave symmetric, the analysis presented above is also extendible to formulate ZVS conditions for switches S_J and S_M.

$$E_{sourced} = \int_{t^*}^{t^* + t_{ZVS}} V_{Z_{TH}}(t) I_{p,TH}(t) \geq 2 C_{OSS,p} V_{p,TH}(t^*) V_{pdc} \tag{6.48}$$

Inverting side approach can be extended to rectifying side as well, where the only variations are the current ($I_{p,TH}$ and I_{pdc}) directions which are reversed compared to the inverting side and accordingly, the corresponding ZVS equations are derived.

As for switching loss calculation, for $I(\zeta) > I_{ZVS,min}$, the turn-on loss and output capacitance loss can be considered 0. If the criterion is not met, the turn-on losses for yth bridge ($y \in [p, s, t]$) are accounted as follows.

$$P_{turnon,y} = 2 V_{ydc} I(\zeta).F_{sw}.t_{on,y} \tag{6.49}$$

where,

$$t_{on,y} = \frac{Q_{gd} R_{g,y}}{V_{gs,y} - V_{pl,y}} + R_{g,y}.C_{in,y}.ln\frac{V_{gs,y} - V_{th,y}}{V_{gs,y} - V_{pl,y}} \tag{6.50}$$

ζ refers to the switching time instant.

Typically, t_{ZVS} is used as the marginal value of the deadtime between complementary switches in a half-bridge because at a higher deadband the additional body-diode conduction loss will occur during the forward bias period. For a selected deadtime of $t_{dt} > t_{ZVS}$, the reverse conduction loss expressed in (6.51) needs to be accounted for.

$$P_{deadtime,y} = 4 V_{f,y} I_y(\zeta)(t_{dt,y} - t_{ZVS,y}).F_{sw} \tag{6.51}$$

The total turn-off loss for semiconductors in yth bridge can be calculated using (6.52–6.53).

$$P_{turnoff,y} = 2 V_{ydc}.I_y(\zeta) F_{sw} t_{off,y} \tag{6.52}$$

where,

$$t_{off,y} = \frac{Q_{gd} R_{g,y}}{V_{pl,y}} + R_{g,y} C_{in,y} \ln\left(\frac{V_{pl,y}}{V_{th,y}}\right) \tag{6.53}$$

Here, $Q_{gd}, R_g, V_{pl}, V_{th}, C_{in}$ represent charge at gate-drain Miller capacitance, gate resistance, plateau voltage, threshold voltage, and input capacitance of the power device, respectively, which can be obtained from device datasheets.

B. **Semiconductor power loss optimization**

A designer may opt in for optimizing a range of cost functions i.e., minimizing RMS currents that might be beneficial for selecting lower current-rated MOSFETs or maximizing the number of ZVS events that would likely improve the EMI performance or the total power loss in the switching network for efficiency maximization. Therefore, in this section, we formulate different types of convex cost functions that are then minimized with respect to the control variables and compared with other non-optimum yet feasible solutions.

In a converter design optimization before choice of the power devices, if one wants to perform conduction loss minimization, a suitable selection of cost function would be (6.54).

$$F_{cond}(F_{sw}, \phi_s, \phi_t) = \sum_{y=p,s,t} I_{y,rms}^2 \qquad (6.54)$$

As for switching loss minimization without any knowledge of power devices to be utilized, a representative cost function would be the product of switching instantaneous current ($I(\zeta)$) and switching frequency (F_{sw}), for a given voltage–power combination, as shown in (6.55). Notably, since $I(\zeta)$ would vary upon different control variable combinations for a given voltage–power combination, our objective here would be to pick the candidate with least current-frequency product.

$$F_{switch}(F_{sw}, \phi_s, \phi_t) = \sum_{y=p,s,t} I_y(\zeta) F_{sw} \qquad (6.55)$$

As for total loss optimization, the cost function would certainly include all possible losses in the switching network per (6.56), each of which can be individually modeled using power device specifications.

$$F_{total}(F_{sw}, \phi_s, \phi_t) = \left(P_{cond} + P_{turnon} + P_{turnoff} + P_{Coss} + P_{deadtime}\right) \qquad (6.56)$$

It is noteworthy that achieving all three objectives of minimizing total semiconductor loss, conduction loss, and switching losses for a given voltage–power combination may not be possible with a single control variable set. Therefore, it is essential to formulate specific optimization functions that provide designers with additional insights during the design and operation of converters. To underscore this point, feasible sets of control variables for extreme voltage–power conditions are listed in Table 6.4. These control variables are selected under the constraint of achieving ZVS across all three ports, while also meeting the required voltage–power conditions at the output ports. The table includes computed values of the aforementioned loss functions, as well as total semiconductor loss for each feasible control variable set. This cataloging facilitates significant observations and reveals trends in losses, aiding in the optimization procedure and an intuitive tuning process. The conduction loss function F_{cond} shows a negative slope with increasing switching frequency which could be explained by the fact that increasing switching frequency and moving closer to the resonant frequency enables the converter to operate with a comparatively lower input impedance phase angle, thus reducing the portion of circulating reactive power in the switching network. For a particular voltage–power condition, control variable with highest switching frequency which is below resonant frequency would demonstrate lowest rms current values. Typically, the current-frequency product cost function (F_{switch}) follows similar trend as F_{cond}, however, it is also impacted by other converter parasitics such as device output capacitance, deadtime, and transformer inter/intra winding capacitance. The total semiconductor loss typically follows the trend by its dominant constituent. Different port voltage and load power conditions are subjected to the optimization routine outlined, and the cost function values at different feasible control variable sets are presented for comparison purpose in Table 6.4. For $V_{pdc} = 400$ V; $V_{sdc} = 600$ V; $V_{tdc} = 28$ V; $P_{sdc} = 750$ W; $P_{tdc} = 500$ W, F_{switch} follows a parabolic curve with the minima achieved at (392 kHz, $- 14.95°$, $- 13.68°$). Due to the switching loss being the most dominating loss type in this case, the same control variable set achieves least total semiconductor loss, albeit a higher value of conduction loss. Taking another test case of $V_{pdc} = 400$ V; $V_{sdc} = 600$ V; $V_{tdc} = 28$ V; $P_{sdc} = 1$ kW; $P_{tdc} = 1$ kW, the control variable with the highest switching frequency yields the least conduction loss primarily due to a greater input impedance magnitude, and least overall semiconductor loss as well. The minimum power loss recorded is 32.63 W. In contrast, an alternative set of control variables performing the same power-voltage combination results in a power loss of 37.07 W. This comparison highlights the efficacy of the loss optimization routine that reduces the overall loss by 13.6%.

6.3.5 Realization and Implementation of Optimized Switching Modulator

As discussed in the previous chapters, look-up table (LUT) based approach is not scalable to higher-order multiport converters due to memory size limitation and hence digital implementation constraints. Therefore, a suitable method would be multivariate polynomial regression-based control variable computation as a function of port voltage gains and load power. Beside from loss optimization, it is a mandatory constraint to ensure required voltage regulations at the output ports. Therefore, a closed-loop control system framework needs to be combined with the loss optimization routine. To address the output voltage regulation on both the ports subjected to line or load transients, a transient detector logic dependent control loop architecture can be formed and is shown in Fig. 6.16. During output load transients, the control variable set generation after being triggered by transient detector logic is transferred from the steady state optimization framework to the closed loop control based on certain logic steps. The sensed values of the output voltage and currents ($V_{sdc}, V_{tdc}, I_{sdc}, I_{tdc}$) are fed to the DSP using appropriate sensor network scaling factors to estimate the instantaneous resistances (R_s^{est}, R_t^{est}) at the output ports. In order to minimize the impact of DC output voltage ripple, current ripple, and sensor delays on load resistance estimation, a 10-point moving average filter (MAF) is implemented. Moving average window length can be selected based on the converter switching frequency

Table 6.4 Feasible control variable sets along with optimized cost functions for different voltage–power combinations with V_{pdc} fixed at 400 V

Voltage–power	Control variable sets	$F_{cond}(A^2)$	$F_{switch}(A.Hz)$	$F_{total}(W)$
600, 28 V, 1, 1 kW	{379 kHz, − 27.3°, − 25.2°}, {370 kHz, − 29.3°, − 27.3°}, {364 kHz, − 30.1°, − 27.4°}, {358 kHz, − 32.2°, − 29.1°}	2722, 2757, 2773, 2788	4.3e6, 6.11e6, 8.2e6, 10.5e6	32.63, 34.08, 35.58, 37.07
600, 28 V, 750, 500 W	{396 kHz, − 14.7°, − 13.4°}, {392 kHz, − 15°, − 13.7°}, {385 kHz, − 15.8°, − 14.6°}, {378 kHz, − 16.3°, − 15°}	693, 714, 782, 842	3.7e6, 3.6e6, 4.2e6, 4.86e6	19.88, 19.62, 19.95, 19.61
550, 25 V, 900, 750 W	{380 kHz, − 20.5°, − 13.6°}, {373 kHz, − 22.8°, − 17.5°}, {367 kHz, − 24.4°, − 17.8°}, {362 kHz, − 25.1°, − 18.6°}	1505, 1601, 1637, 1635	1.37e7, 1.35e7, 1.38e7, 1.49e7	26.56, 29.29, 29.56, 29.98
550, 25 V, 750, 250 W	{402 kHz, − 6.5°, − 2.9°}, {396 kHz, − 7.1°, − 2.8°}, {388 kHz, − 8.1°, − 3.49°}, {380 kHz, − 9.3°, − 2.9°}, {374 kHz, − 10.1°, − 5.4°}	242, 260, 274, 264, 281	1.32e7, 1.25e7, 1.35e7, 1.33e7, 1.35e7	21.21, 21.29, 21.219, 21.3, 21.33
525, 22 V, 750, 250 W	{406 kHz, − 5.1°, − 2.6°}, {398 kHz, − 6°, − 3.3°}, {390 kHz, − 6.4°, − 3.8°}, {384 kHz, − 7.6°, − 4.6°}, {382 kHz, − 8.3°, − 6.6°}, {374 kHz, − 10°, − 8.3°}	603, 710, 830, 888, 822, 756	1.6e7, 1.8e7, 1.94e7, 2.0e7, 1.95e7, 1.87e7	21.15, 22.12, 22.61, 22.87, 23.2, 23.05

and ripple frequency components at the output port voltages. The ratios of averaged resistance values with the previously stored values are calculated at every DSP execution cycle to detect whether or not any load transients have occurred at any port. If the measured variations of the ratios are found to be above a certain threshold, the transient detector triggers and the converter switches over to the closed loop control framework by setting the voltage regulation as the first priority. The PI controllers output the phase shifts, while the frequency is obtained from the phase-frequency contour that can be coded within the DSP as a polynomial regression as a function of port gains and load power. After termination of the transient event, once the relative variations in port resistances are found to be within ±5% margin, the DSP switches back to the steady-state optimization framework for loss minimization after a delay of 100–200 ms which is conservatively selected

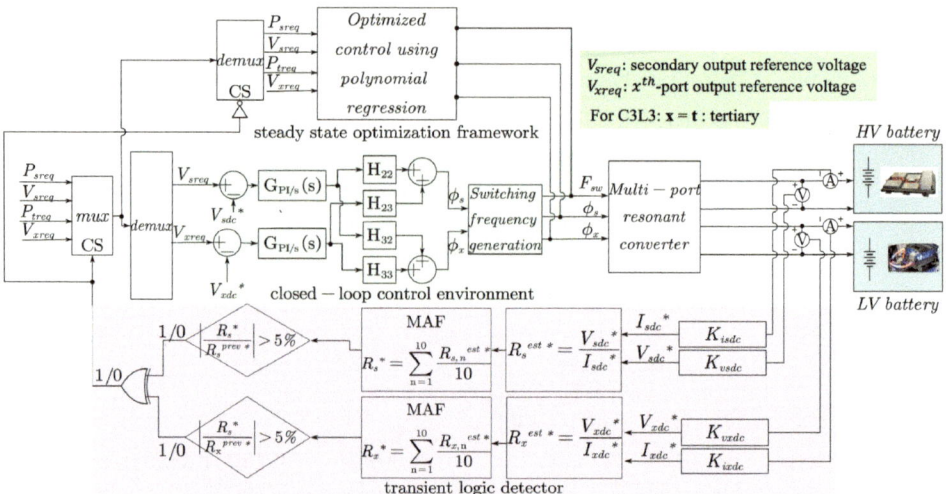

Fig. 6.16 Transient detector logic based closed loop control and optimization framework for three port C3L3 resonant converter

for the converter output to reach the steady state for a control bandwidth of ~5–10 kHz. During periods of load transients, there is a chance that the converter may operate with a suboptimal control variable set for the desired output voltage and power. Nevertheless, this effect on the converter's continuous operational efficiency is minimal because the transient intervals are extremely brief compared to the steady-state periods. The primary benefit of this logic-triggered closed-loop system, implemented within the DSP as an interrupt to the main loop, is that it ensures the converter consistently operates at the most optimized control variable settings. Additionally, it demonstrates robustness against the frequent output load variations.

The steady-state optimization framework applied to the C3L3 resonant converter, illustrated in Fig. 6.17, is executed through a blend of offline and online computations. The control variables, optimized for minimal total loss across all voltage–power scenarios illustrated in Table 6.4, are used as inputs. Additionally, parameters related to the converter's semiconductor parasitics are provided to a high-performance offline computer for formulation of the loss objective functions. For each voltage–power combination, multiple output solution sets are available. An optimization routine, constrained by specific power transfer criteria, is used to determine the most optimal control variable from these sets. By selecting a few pre-defined voltage–power combinations within the converter's operational range, the mathematical model generates optimal control variable sets for each combination, which are then compiled and organized into a table. These voltage–power pairs and their corresponding optimal control variables are fed into a polynomial regression algorithm. This algorithm, executed offline on a computing machine using a least

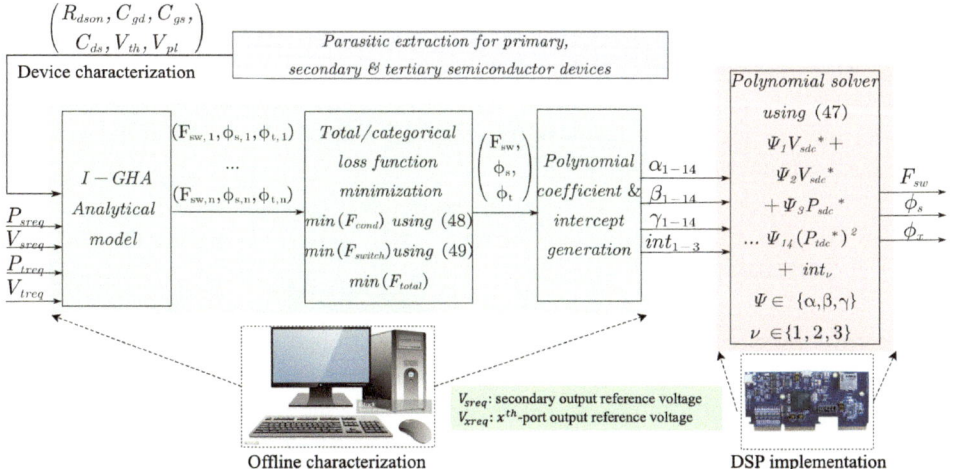

Fig. 6.17 Hybrid offline-online polynomial regression based steady-state optimization framework for three port C3L3 resonant converter

mean square error (LMSE) estimator, calculates the polynomial coefficients and intercepts. The MATLAB Polynomial Regression toolbox can be employed to manage the extensive computations required for generating these coefficients and intercepts. The generation of polynomial coefficients and intercepts marks the end of the offline phase of this control architecture. This is followed by the online phase, where these polynomial coefficients and intercepts for each of the three control variables are programmed into the read-only memory (ROM) of the DSP. The computation time for this process is directly related to the order of the polynomial function used in the model.

If the computational time for determining the duty parameters exceeds the converter's switching cycle, the control loop will be updated at a reduced rate compared to the switching rate and the sampling rate of the sensed voltage and current signals. This mismatch may result in signal aliasing and potential instability in the control loop, particularly in case of load or line transients. For the DSP, computing control variables using polynomial regression involves a sequence of mathematical operations, such as multiplications and additions, executed through specific instructions. The total time required to perform these operations is the sum of the times needed to execute each individual instruction. Implementing a second-order polynomial requires 28 multiplications and 14 additions for each control variable. With three control variables for C3L3 as per Fig. 6.16, this results in a total of 84 multiplications and 42 additions.

An example formulation of a control variable is presented in (6.57):

$$F_{sw} = \alpha_1 V_{sdc} + \alpha_2 V_{tdc} + \alpha_3 P_{sdc} + \alpha_4 P_{tdc} + \alpha_5 V_{sdc} V_{tdc}$$
$$+ \alpha_6 P_{sdc} P_{tdc} + \alpha_7 V_{sdc} P_{sdc} + \alpha_8 V_{sdc} P_{tdc} + \alpha_9 V_{tdc} P_{sdc} + \alpha_{10} V_{tdc} P_{tdc}$$

$$+ \alpha_{11} V_{sdc}^2 + \alpha_{12} V_{tdc}^2 + \alpha_{13} P_{sdc}^2 + \alpha_{14} P_{tdc}^2 + int_1 \qquad (6.57)$$

where, α_1 to α_{14} are the polynomial coefficients and int_1 is the intercept. Additionally, similar equations can be formulated for other control variables ϕ_s and ϕ_t with coefficients $\beta_1 - \beta_{14}$ and $\gamma_1 - \gamma_{14}$ and intercepts of int_2 and int_3, respectively, as part of the steadystate optimization framework in Fig. 6.17.

6.4   Conclusions

To improve the effectiveness of traditional PWM-controlled multi-port converters across a broad range of gain and port powers, this chapter investigates a multi-variable loss optimization approach for a three-port resonant topology—the triple-active C3L3 converter. Furthermore, a semiconductor loss model for a three-port resonant converter is developed, followed by a detailed analysis of categorical and overall loss optimization. This analysis integrates a hybrid modulation scheme involving switching frequency and inter-bridge phase shifts. The three-port resonant converter operates within a steady-state optimization framework, where the switching patterns encode information on switching frequency and inter-bridge phase shifts. The control variable set synthesis routine is implemented in the controller using a polynomial regression approach that helps the converter achieve ZVS and net switching network loss minimization while reducing the full-bridge RMS currents under extreme operating conditions. Based on the I-GHA model, the resultant port currents exhibit a mean variation of $\leq 3\%$ in peak and RMS values compared to the real-time hardware-acquired port current, and a mean variation of $\leq 10\%$ in switching instant values. While the mean variation in switching instant values may be higher in certain scenarios, the overall impact on the total semiconductor loss estimation remains at $\leq 5\%$, falling within an acceptable threshold for a multi-port resonant converter modeling analysis. This level of variation is significantly lower than that observed with current state-of-the-art modeling techniques like GHA and FHA, which typically show an error of $\geq 10\%$ in total semiconductor loss estimation.

References

1. A. Chandwani, A. Mallik and A. Akturk, "Steady-State Model Derived Multi-variable Loss Optimization for Triple Active C3L3 Resonant Converter," in IEEE Transactions on Transportation Electrification, https://doi.org/10.1109/TTE.2023. 3266744.
2. A. Sankar, A. Mallik and A. Khaligh, "Extended Harmonics Based Phase Tracking for Synchronous Rectification in CLLC Converters," in IEEE Transactions on Industrial Electronics, vol. 66, no. 8, pp. 6592-6603, Aug. 2019, https://doi.org/10.1109/TIE.2018.2874348
3. Q. Shang, F. Xiao, Y. Fan, R. Wang, H. Qin and T. Song, "Parameter Estimation of DAB Converter Using Intelligent Algorithms and Steady-State Modeling Considering Nonidealities," in IEEE Transactions on Industrial Electronics, https://doi.org/10.1109/TIE.2024.3390731.

4. J. -H. Jung, H. -S. Kim, M. -H. Ryu and J. -W. Baek, "Design Methodology of Bidirectional CLLC Resonant Converter for High- Frequency Isolation of DC Distribution Systems," in IEEE Transactions on Power Electronics, vol. 28, no. 4, pp. 1741-1755, April 2013, https://doi.org/10.1109/TPEL.2012.2213346.

5. P. He and A. Khaligh, "Comprehensive Analyses and Comparison of 1 kW Isolated DC–DC Converters for Bidirectional EV Charging Systems," in IEEE Transactions on Transportation Electrification, vol. 3, no. 1, pp.147-156, March 2017, https://doi.org/10.1109/TTE.2016.2630927.

6. X. Li, J. Huang, Y. Ma, X. Wang, J. Yang and X. Wu, "Unified Modeling, Analysis, and Design of Isolated Bidirectional CLLC Resonant DC–DC Converters," in IEEE Journal of Emerging and Selected Topics in Power Electronics, vol. 10, no. 2, pp. 2305-2318, April 2022, https://doi.org/10.1109/JESTPE.2022.3145817.

7. S. Mungekar and A. Mallik, "An Improved GHA-Enabled Steady State Model-Derived Semiconductor Loss Optimization for a Three-Port C3L3 Resonant Converter," in IEEE Transactions on Power Electronics, vol. 39, no. 6, pp. 7654-7674, June 2024, https://doi.org/10.1109/TPEL.2024.3373514.

Refined Modeling of Two-Port Resonant and Non-resonant MABs Including Transformer Parasitic Non-idealities and Deadtime Effects

7

7.1 Introduction to Transformer Capacitive Non-idealities

In the previous chapters, the modeling of MABs has been carried out considering only resistive non-idealities present in the converter network. However, in practice, the operation and performance of any isolated converter would have considerable impacts arising from capacitive non-idealities present in the transformer. For example, a simplified equivalent of a two-winding transformer would have intra-winding capacitances across the primary and secondary windings, and inter-winding capacitance between primary and secondary windings [1]. To be more generalized, for an N-winding transformer [2], its simplified equivalent would have N intra-winding capacitances and $\binom{N}{2}$ number of inter-winding capacitances. Therefore, in the network modeling of an N-port MAB, inclusiveness of the capacitance effect would add $\frac{N(N+1)}{2}$ extra order to the basic system with no capacitive impacts considered. More important, such capacitive effects become more prominent in case of planar transformer configuration [3] where the overlap areas between any successive layers of turns are typically larger than a wire-wound transformer with Litz winding or hook-up wire-based winding. Also, the PCB material, which could be FR4, or Teflon or Alumina based on the type of application, would have considerable dielectric constants that further contribute towards forming sizable inter-turn overlap capacitance. As a reference, the relative permittivity values of some common PCB dielectrics are mentioned below:

$$\epsilon_r = 3.8 \, to \, 4.8 \text{ (for FR4); } \epsilon_r = 8 \, to \, 10 \text{ (for Alumina); } \epsilon_r \approx 2 \text{ (for PTFE).}$$

© The Author(s), under exclusive license to Springer Nature Switzerland AG 2025
A. Mallik and S. Dey, *Switching Modulator Optimization in Isolated Power Converters*,
Synthesis Lectures on Power Electronics, https://doi.org/10.1007/978-3-031-81576-8_7

For a geometrical representation, a non-interleaved planar transformer with 22:1 turns ratio for a high step-down isolated converter [4] is shown in Fig. 7.1. There are two 4-layer PCBs, one containing 22 primary turns in 7-4-4-7 orientation and the other containing 1 high-current secondary turn in 1-1-1-1 (all parallel) configuration. The insulation between the copper layers is provided by the prepreg and core layers, which have thicknesses of h_{pr} and h_c, respectively. Additionally, the transition from one layer to the next is achieved through conventional vias in the PCB, with their hole sizes designed to meet the current-carrying needs of the windings according to IPC 2221 standards. Assuming that the potential across the winding varies linearly with the turns, the potential at each primary turn can be expressed as follows:

$$V_y = \frac{(n+1) - y}{n} V_p; \ y \in \{1, 2, 3 \dots, n\} \text{ for } y \in \{1, 2, 3 \dots, n\} \tag{7.1}$$

where, n is the number of turns and V_p is the primary winding voltage excitation. Therefore, a voltage gradient ($V_{y,z}$) exists between the two adjacent windings and the windings in two successive layers, which essentially leads to the formation of virtual capacitors. The overlapping capacitance between two adjacent conductors can be formulated by the conductor overlap area ($S_{y,z}$) between turns y and z and spacing between them (d), as follows.

$$C_{y,z} = \frac{\epsilon_0 \epsilon_r S_{y,z}}{d} \tag{7.2}$$

$$S_{y,z} = \int_0^{l_t} w_0 dl \tag{7.3}$$

where ϵ_0 and ϵ_r denote the free-space permittivity and dielectric constant of the medium, and w_0 is the overlapping conductor width and dl represents an infinitesimally small sectional length of a turn, which is integrated over the entire circumference to form a complete turn of length l_t.

Referring to Fig. 7.1, since the overlap area $\left(\int_0^{l_t} h_1 dl\right)$ between the two turns is very small with air (distance between the conductors: h_{is}) being the dielectric medium between them, the turn-to-turn capacitance is negligible, and thus, its effect can be ignored. On the other hand, the capacitance between adjacent layers can be formulated by analyzing the total energy associated with the electric field between the two layers.

$$E_l = \sum_{l=1}^{l_n} \frac{1}{2} [C_{y,z} V_{y,z}^2]_{layer} \tag{7.4}$$

$$E_{l_t} = \sum_{l=1}^{l_n} E_l \tag{7.5}$$

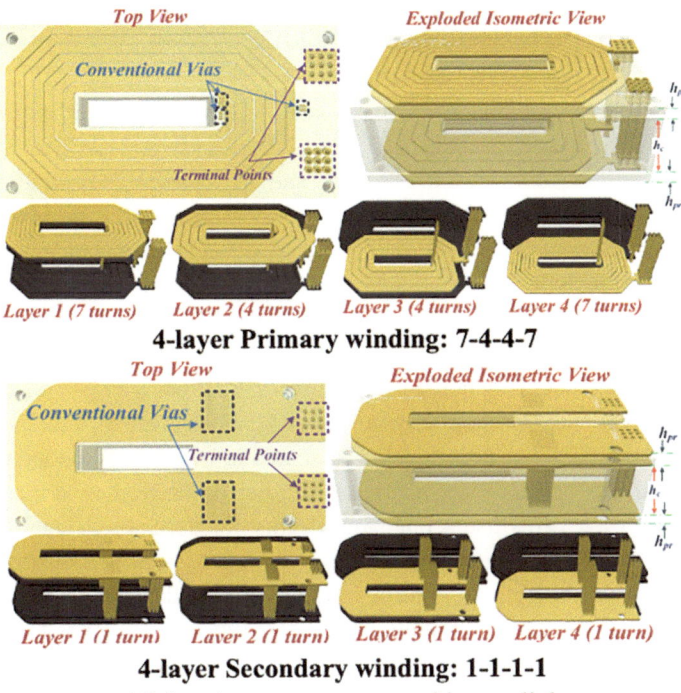

4-layer Primary winding: 7-4-4-7

4-layer Secondary winding: 1-1-1-1

All four layers are connected in parallel

Fig. 7.1 PCB winding arrangement for non-interleaved {[7P-4P-4P-7P], [1S∥1S∥1S∥1S]} configuration

where, $V_{y,z}$ is the potential difference between two conductor surfaces, and E_l denotes the total energy in a layer l. Using the above equations, the effective inter or intra-winding capacitance can be formulated as:

$$C_{in} = \sum_{t=1}^{l_t\,(no.\,of\,layers)} \frac{2E_{l,t}}{V_{y,z}^2} \tag{7.6}$$

There are several works [4–7] in the literature that lay the foundation of transformer parasitic capacitance derivation methods and present some specific testcases of converters such as LLC, CLLC, and DAB.

7.2 Tank Current Reconstruction Model for a Capacitance Parasitic-Inclusive MAB Network

An equivalent network schematic for an isolated MAB including all possible inter- and intra-winding capacitances is shown in Fig. 7.2. C_k $\forall k \in [1, N]$ defines the equivalent intra-winding capacitance on kth port, and C_{ij} defines inter-winding capacitance between ith and jth port windings of the transformer. L_{li} denotes the leakage inductance of the ith port winding of the transformer, while Z_i denotes the equivalent tank impedance of ith port, which include lumped equivalent of all series resistive non-idealities. These non-idealities consist of FET on-state resistances, transformer AC winding resistance, and any capacitor ESR present. Keeping the port-1 as the reference bridge, the phase angle for the kth bridge voltage is denoted as $-\phi_k$.

For the sake of analysis simplicity, a 2-port MAB equivalent, inclusive of resistive, inductive, and capacitive non-idealities, is presented in Fig. 7.3. To understand the impact of parasitic elements on converter operation and performance, it is critical to reproduce the tank current waveforms using our Non-ideality Inclusive Frequency Domain Model (NIFDM) based on GHA and compare them against ideal circuit waveforms. To assess the performance impact, the optimal control variables can be determined using this comprehensive modeling approach. By comparing the cost functions, such as total power losses, between scenarios that include the updated control variable set and those that exclude capacitance parasitics, we can evaluate the differences for a given converter specification. In this section, the current waveforms are analytically derived, reconstructed, and compared for a range of capacitance parasitic magnitudes.

Here, Z_p and Z_s are the series equivalents of tank impedances and resistive non-idealities on the primary and secondary sides, respectively. The characteristics of Z_p and Z_s would vary depending on the circuit topologies, which are detailed on a case-by-case basis in the following section. In Fig. 7.3, C_{ip}, C_{is}, C_{ps} denote the primary intra-winding, secondary intra-winding, and primary-secondary inter-winding capacitances, respectively. L_{lp} and L_{ls} represent the primary and secondary leakage inductances of the transformer. Further, the ac winding resistances of the transformer's primary and secondary windings are denoted as R_{lp} and R_{ls}. The network in Fig. 7.3b is synthesized by performing a star-delta transformation of the Y-type resistive-inductive network in Fig. 7.3a. The important equations corresponding to the networks in Fig. 7.3 are presented below.

For a CLLC-type network:

$$Z_p = \left(R_p + j\omega L_p - j\frac{1}{\omega C_p} \right) \tag{7.7}$$

$$Z_s = \left(R_s + j\omega L_s - j\frac{1}{\omega C_s} \right) \tag{7.8}$$

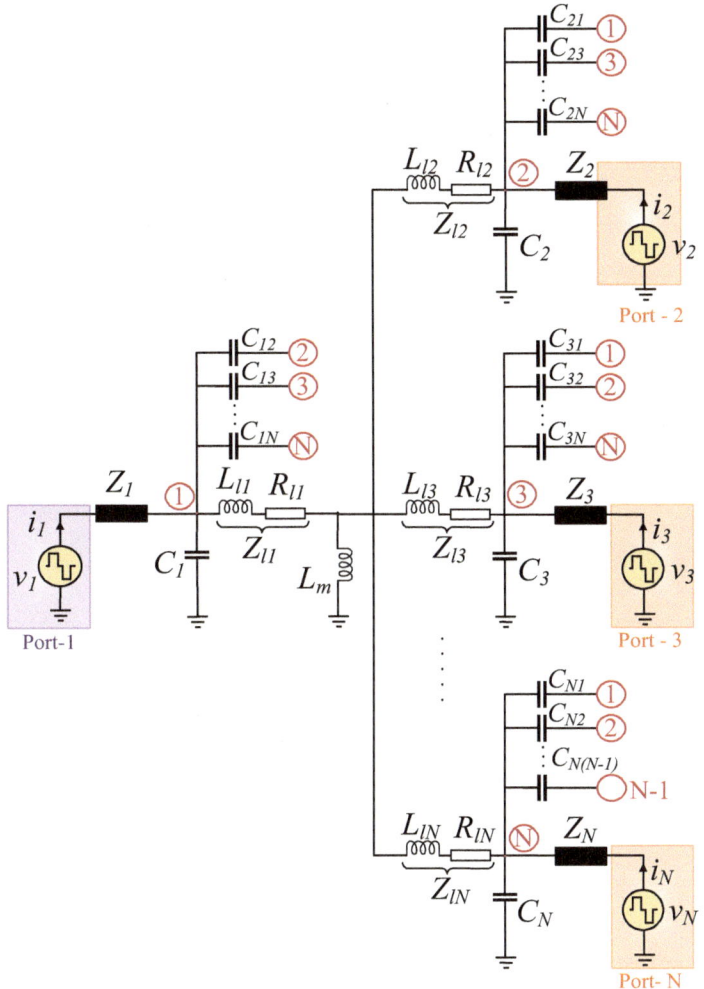

Fig. 7.2 MAB equivalent network incorporating inductive and capacitive parastics arising from the isolation transformer

For an LLC-type network, Z_p is the same as above; however, secondary side tank does not include any power transfer impedance, so Z_s would only consist of the lumped equivalent of trace resistance and FET on-state resistance ($R_{on,s}$). For a full-bridge secondary, Z_s would be as follows.

$$Z_s = R_s = R_{trace,s} + 2R_{on,s} \tag{7.9}$$

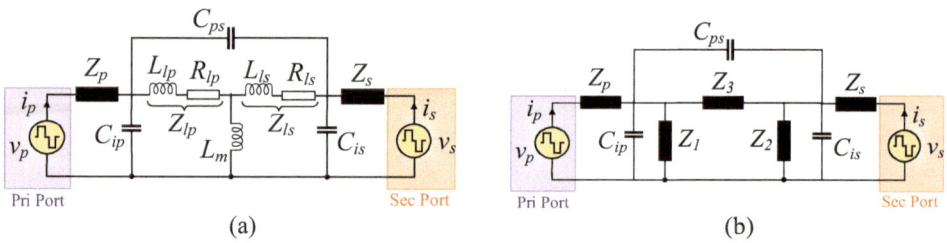

Fig. 7.3 **a** Two-port MAB equivalent network including R-L-C parasitics, **b** star-delta converted version of the parent network in (**a**)

For a DAB-type network, even though there might be DC blocking capacitors on either primary or secondary or both sides, their impedances at switching frequency are negligibly low compared to those of power transfer inductances. Therefore, the following can be written.

$$Z_p = \left(R_p + j\omega L_p\right) \tag{7.10}$$

$$Z_s = (R_s + j\omega L_s) \tag{7.11}$$

According to the network in Fig. 7.3b, the transformed inductances are expressed as follows.

$$Z_1 = Z_{lp} + Z_{lm} + \frac{Z_{lp}Z_{lm}}{Z_{ls}} \tag{7.12}$$

$$Z_2 = Z_{ls} + Z_{lm} + \frac{Z_{ls}Z_{lm}}{Z_{lp}} \tag{7.13}$$

$$Z_3 = Z_{lp} + Z_{ls} + \frac{Z_{lp}Z_{ls}}{Z_{lm}} \tag{7.14}$$

Here, $Z_{lp} = R_{lp} + j\omega L_{lp}$; $Z_{ls} = R_{ls} + j\omega L_{ls}$; $Z_{lm} = j\omega L_m$; and $\omega = 2\pi f_{sw}$, with f_{sw} being the converter switching frequency.

For DAB-like converters, typically the magnetizing inductance is designed to be much greater than transformer leakage inductances, i.e., $L_m \gg L_{lp}$; $L_m \gg L_{ls}$.

Therefore, the following approximation can be made for a DAB: $Z_1 = Z_2 \approx 2Z_{lm}$ and $Z_3 \approx 2Z_{lp}$.

However, for resonant converters like CLLC, LLC, CLL, LCC, the above approximation cannot be assumed. The shunt combinations of Z–C pairs such as $\left(Z_1, C_{ip}\right)$, (Z_2, C_{is}), (Z_3, C_{ps}) can be condensed to form the further simplified network shown in Fig. 7.4a, which then after undergoing a delta-star transformation of $\left(Z_1', Z_2', Z_3'\right)$ network produces the network in Fig. 7.4b.

Fig. 7.4 **a** Equivalent 2-port generalized parasitic-inclusive MAB network. **b** Simplified T-network equivalent of 2-port generalized parasitic-inclusive MAB circuit

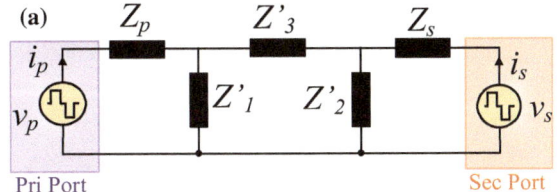

(a)

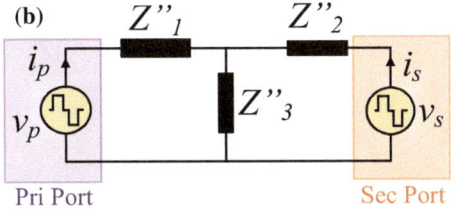

(b)

Z_1', Z_2', Z_3' in Fig. 7.4a can be expressed as follows.

$$Z_1' = Z_1 \| \frac{1}{j\omega C_{ip}} \tag{7.15}$$

$$Z_2' = Z_2 \| \frac{1}{j\omega C_{is}} \tag{7.16}$$

$$Z_{3'} = Z_3 \| \frac{1}{j\omega C_{ps}} \tag{7.17}$$

The further transformed T-network impedances Z_1'', Z_2'', Z_3'' in Fig. 7.4b are formulated in (7.18–7.20).

$$Z_1'' = Z_p + \frac{Z_1' Z_3'}{Z_1' + Z_2' + Z_3'} \tag{7.18}$$

$$Z_2'' = Z_s' + \frac{Z_2' Z_3'}{Z_1' + Z_2' + Z_3'} \tag{7.19}$$

$$Z_3'' = \frac{Z_1' Z_2'}{Z_1' + Z_2' + Z_3'} \tag{7.20}$$

Now, the individual harmonic components of the primary and secondary tank currents can be derived using superposition theorem on this linear equivalent network. The kth harmonic (i.e., $\omega = k\omega_s = 2\pi k f_{sw}$) components of i_p and i_s are expressed below, where $V_{p,k}$ and $V_{s,k}$ denote kth harmonic amplitude of the primary and secondary bridge voltages, respectively, which are: $V_{p,k} = \frac{4V_{pdc}}{k\pi}cos(k\delta_p)$ and $V_{s,k} = \frac{4V_{sdc}}{k\pi}cos(k\delta_s)$, where V_{pdc} and V_{sdc} are the DC link voltages of the primary and secondary bridges.

$$\widetilde{i_{p,k}} = \frac{V_{p,k}\angle 0°}{Z_1''(k\omega_s) + Z_2''(k\omega_s)||Z_3''(k\omega_s)} - \frac{V_{s,k}\angle - k\phi_s}{Z_2''(k\omega_s) + Z_1''(k\omega_s)||Z_3''(k\omega_s)}$$
$$\cdot \frac{Z_3''(k\omega_s)}{Z_1''(k\omega_s) + Z_3''(k\omega_s)} \tag{7.21}$$

$$\widetilde{i_{s,k}} = -\frac{V_{p,k}\angle 0°}{Z_1''(k\omega_s) + Z_2''(k\omega_s)||Z_3''(k\omega_s)} \cdot \frac{Z_3''(k\omega_s)}{Z_2''(k\omega_s) + Z_3''(k\omega_s)}$$
$$+ \frac{V_{s,k}\angle - k\phi_s}{Z_2''(k\omega_s) + Z_1''(k\omega_s)||Z_3''(k\omega_s)} \tag{7.22}$$

The net primary and secondary tank currents can be formulated as the phasor summation of their all harmonics, as per (7.23)–(7.24), considering m number of harmonics.

$$i_p = \sum_{k=1,3,5...}^{2m+1} \widetilde{i_{p,k}} \tag{7.23}$$

$$i_s = \sum_{k=1,3,5...}^{2m+1} \widetilde{i_{s,k}} \tag{7.24}$$

In general, kth harmonics for primary and secondary tank currents and their net RMS currents can be expressed in time domain in (7.25)–(7.28), where $\theta_{k,ip}$ and $\theta_{k,is}$ represent the angles of $\widetilde{i_{p,k}}$ and $\widetilde{i_{s,k}}$ phasors.

$$\widetilde{i_{p,k}} = A_{k,ip}\sin(k\omega_s t + \theta_{k,ip}) \tag{7.25}$$

$$\widetilde{i_{s,k}} = B_{k,is}\sin(k\omega_s t + \theta_{k,is}) \tag{7.26}$$

$$I_{p,RMS} = \sqrt{\frac{1}{2}\sum_{k=1,3,...}^{\infty} A_{k,ip}^2} \tag{7.27}$$

$$I_{s,RMS} = \sqrt{\frac{1}{2}\sum_{k=1,3,...}^{\infty} B_{k,is}^2} \tag{7.28}$$

The total AC power fed from the primary AC bridge and sunk at the secondary AC bridge can be represented as the sum of individual active power contained by each harmonic, shown as follows.

Primary AC power supplied: $P_{p,ac} = \dfrac{4V_{pdc}}{\pi}\displaystyle\sum_{k=1,3...}^{\infty} \dfrac{A_{k,ip}}{k}\cos(\theta_{k,ip})$ \hfill (7.29)

$$\text{Secondary AC power sunk: } P_{s,ac} = \frac{4V_{sdc}}{\pi} \sum_{k=1,3...}^{\infty} \frac{B_{k,is}}{k} \cos\left(\theta_{k,is} + k\phi_s\right) \quad (7.30)$$

Now let us take two case studies, one with CLLC converter and the other with DAB converter and reconstruct the tank current waveforms and examine the impact of different values of transformer parasitic elements.

- **CLLC case study**: Assume, the converter is operating at 600–28 V conversion with a transformer turns ratio of 21:1 and is designed to have a resonant frequency of 500 kHz, with the following parameters, all reflected on the primary side.

$$L_p = 8.1\,\mu\text{H}; \ C_p = 10\,\text{nF}; \ L_m = 120\,\mu\text{H}; \ L_s = 8.1\,\mu\text{H}; \ C_s = 10\,\text{nF};$$

The resonant tank parameters of the CLLC converter mentioned above are referred to the primary side of the transformer. Further, the lumped resistance equivalents and the transformer winding resistances on the primary and secondary (reflected to primary) are: $R_p = 61$ mΩ; $R_s = 700$ mΩ; $R_{lp} = 387$ mΩ; $R_{ls} = 472$ mΩ and the leakage inductances are as follows: $L_{lp} = 2\,\mu\text{H}, L_{ls} = 2\,\mu\text{H}$.

Figure 7.5a, b shows the CLLC primary and secondary bridge voltage, as well as the bridge or tank current waveforms in the base case or case-1 where no capacitive parasitics are considered. Table 7.1 showcases the control and circuit parameters under test and the measured power flow and winding current RMS values. With a switching frequency of 460 kHz and a secondary bridge voltage phase shift of 0.15 rad, an output power of 3.63 kW is achieved with 3.7 kW of power fed by the primary side, accounting for losses due to non-ideal resistive elements.

To study the impact of capacitive parasitics on winding currents, intra-winding and inter-winding capacitances were varied separately. First, by keeping $C_{ps} = 20$ pF, and $C_{is} = 50$ pF, C_{ip} is varied from 10–100 pF. The resulting i_p and i_s waveforms are illustrated in Fig. 7.6a, and the results are quantified in Table 7.1 under case-2 and case-3. In the second round, C_{ip} and C_{is} were kept at their nominal values of 100pF and 70pF, respectively, while C_{ps} was varied from 20 to 1000 pF. This impact of inter-winding capacitance on the tank current waveforms is shown in Fig. 7.6b, with results captured in 7.1 under case-4 and 5.

It is to be kept in mind by referring to Fig. 7.2a that the primary tank current in the CLLC converter is actually divided into three components – one flowing through the transformer winding that is responsible for main power transfer and two leaky components flowing through the intra-winding capacitance (C_p) and inter-winding feedback capacitance (C_{ps}) branches. Thus, the resonating components from all possible L-C combinations in leaky paths appear in the main tank currents, i_p and i_s. Observing current

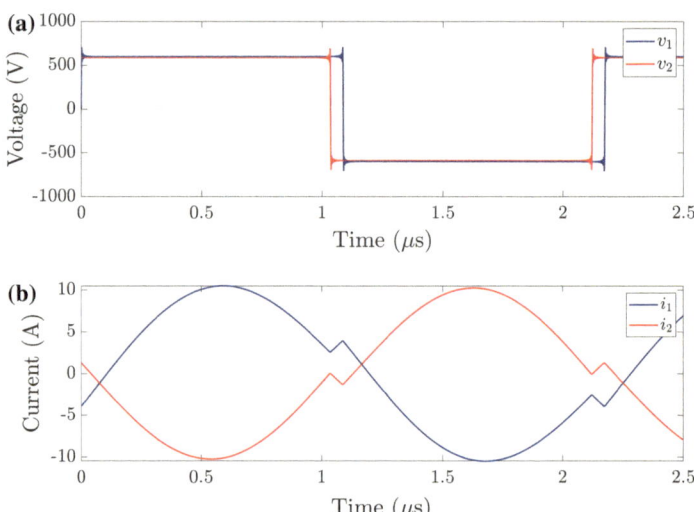

Fig. 7.5 a Bridge Voltages of the CLLC Converter with 600–28 V voltage conversion and $f_{sw} = 460$ kHz. **b.** CLLC Winding Currents without considering any transformer parasitics (here, $i_1 = i_p$ and $i_2 = i_s$; $v_1 = v_p$ and $v_2 = v_s$)

plots of Fig. 7.6a for case-2 with $C_{ip} = 10$ pF and $L_p = 8.1$ µH reveals an 18 MHz high-frequency AC component superimposed on the 460 kHz average ac current i_p. This arises from resonance between C_{ip} and L_p at a resonant frequency of $\frac{1}{2\pi\sqrt{L_pC_{ip}}} = 17.7$ MHz. As C_{ip} increases to 100 pF in case-2, the resonant frequency drops to 5.59 MHz, causing high-frequency oscillation near 5.6 MHz. A similar effect is observed on the secondary winding current i_s. Although L_s and C_{is} remain unchanged between case-2 and 3, the frequency of the resonance between C_{ip} and $(L_s + L_{lp} + L_{ls})$ is changed due to change in C_{ip}. This explains the effect of varying primary side intra-winding capacitance on the secondary winding current. Another important point to note in Fig. 7.6a is how the amplitude of high-frequency parasitic resonant components varies with changes in C_{ip}. As C_{ip} increases, the effective quality factor of the parasitic resonant path decreases, leading to a lower characteristic impedance. That naturally results in an increase in the amplitudes of parasitic resonant components in such a voltage-driven network. The green curve in Fig. 7.6a, which has a C_{ip} ten-times higher than that in the red curve, shows nearly three-times greater amplitudes of the high-frequency overriding components due to $L_p - C_{ip}$ resonant oscillations.

Another interesting perspective to note is that as the parasitic intra-winding capacitances increases, their corresponding resonant frequencies with the primary and secondary inductances decrease, thus reducing the switching frequency harmonic order coinciding with the parasitic resonant frequency. Consequently, lower-order harmonics have higher voltage amplitudes, and minimal resonance impedance paths result in higher current

Table 7.1 CLLC converter operation case studies with different transformer parasitics

CLLC converter parameters	Case 1	Case 2	Case 3	Case 4	Case 5
	Benchmark without any capacitive parasitics	C_{ip} variation with constant C_{is} and C_{ps}		C_{ps} variation with constant C_{ip} and C_{is}	
V_{pdc}/V'_{sdc}	600 V/28 V				
L_p/L_s	10.1 μH/10.1 μH				
L_{lp}/L_{ls}	2 μH/2 μH				
R_p/R_s	61 mΩ/700 mΩ				
R_{lp}/R_{ls}	387 mΩ/472 mΩ				
C_p/C_s	10 nF/10 nF				
f_{sw}/f_r	460 kHz/500 kHz				
ϕ_s	−0.15 rad				
C_{ip}	~0 pF	10 pF	100 pF	100 pF	100 pF
C_{is}	~0 pF	50 pF	50 pF	70 pF	70 pF
C_{ps}	~0 pF	20 pF	20 pF	20 pF	1000 pF
$i_{p,RMS}/i_{s,RMS}$	7.20A/6.91 A	7.145 A/ 7.175 A	7.18 A/ 7.197 A	7.39 A/ 7.40 A	7.516 A/ 7.552 A
$P_{p,ac}/P_{s,ac}$	3.7 kW/ 3.63 kW	3.68 kW/ 3.61 kW	3.64 kW/ 3.57 kW	3.72 kW/ 3.64 kW	3.78 kW/ 3.7 kW

amplitudes. Thus, higher C_{ip} or C_{is} lead to higher amplitude of the high-frequency current ringing.

By observing Fig. 7.6b, it is clear that there is no direct correlation between C_{ps} and the tank currents' ringing frequency or amplitude. This lack of correlation arises because the leakage currents flow predominantly through the intra-winding capacitors, which primarily determine the tank current's shape. It is also noticeable that the shapes of the winding currents, their RMS values ($i_{p,RMS}$, $i_{s,RMS}$), and the input and output AC power of the converter network vary minimally across different test cases with varying transformer parasitics. The impact of parasitic components on the converter's power flow becomes more prominent when the parasitic capacitances are significant relative to the resonant capacitances. In such cases, the parasitic components actively participate in the power flow. Therefore, for high-frequency CLLC converters operating at or above 500kHz, including the transformer non-idealities in the converter modeling becomes essential. Figures 7.5 and 7.6 further illustrate that the instantaneous winding current values during switching instances are significantly influenced by parasitic elements. Consequently, modeling or investigating the soft-switching performance of the converter using an ideal model can be misleading and may result in non-optimal converter operation.

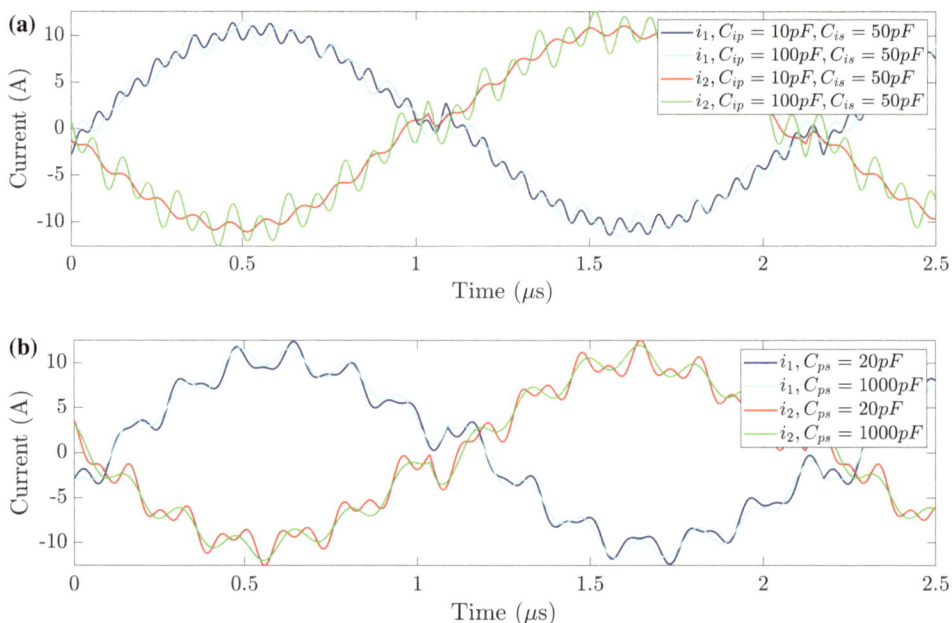

Fig. 7.6 a CLLC Winding Currents with varying primary side intra-winding capacitance. (here, $i_1 = i_p$ and $i_2 = i_s$). **b** CLLC Winding Currents with varying inter-winding capacitance. (here, $i_1 = i_p$ and $i_2 = i_s$)

DAB Case Study: Consider a Dual Active Bridge (DAB) converter designed for a 160–100 V conversion at a switching frequency of 100 kHz, with a transformer turns ratio of 7:5. The converter's parameters are as follows: Primary side inductor, $L_p = 5.96\,\mu H$; secondary side inductor $L_s' = 2.85\,\mu H$ (with $L_s = L_s' \times \left(\frac{7}{5}\right)^2$); primary side dc blocking capacitor $C_p = 10\,\mu F$; and secondary side dc blocking capacitor $C_s' = 14\,\mu F$. The primary-referred transformer winding resistances and the lumped resistance equivalents, including MOSFET on-state resistances, DC blocking capacitor ESR, and inductor AC resistances, are: $R_p = 131 m\Omega$; $R_s = 117\ m\Omega$; $R_{lp} = 210\ m\Omega$; $R_{ls} = 200\ m\Omega$ and the leakage inductances are as follows: $L_{lp} = 0.24\,\mu H$, $L_{ls} = 0.24\,\mu H$.

In this study, the primary and secondary inductor currents of the DAB converter are analyzed considering different transformer parasitic parameters, as shown in Table 7.2. The PWM control parameters are kept constant: $\{\phi_s, \delta_p, \delta_s\} = \{0.13,\ 0.72,\ 0.52\ \text{rad}\}$. Figure 7.7 depicts the base case voltage and current plots without capacitive parasitics, showing an input ac power of 218 W and an output ac power of 214 W, considering resistive losses in the circuit.

To study the impact of transformer parasitics on the winding currents, the intra-winding and inter-winding transformer capacitances are varied separately. Under test case-2, the introduced non-idealities are $C_{ps} = 1.1\,\text{nF}$, $C_{ip} = 50\,\text{pF}$, and $C_{is} = 60\,\text{pF}$. Figure 7.8a

Table 7.2 DAB converter operation case studies with different transformer parasitics

DAB Converter Parameters	Case 1	Case 2	Case 3	Case 4	Case 5
	Benchmark without any Capacitive Parasitics	Simultaneous C_{ip}, C_{is} variation with constant C_{ps}		C_{ps} variation with constant C_{ip} and C_{is}	
V_{pdc}/V'_{sdc}	160 V/100 V				
L_p/L_s	5.96 µH/5.586 µH				
L_{lp}/L_{ls}	0.24 µH/0.24 µH				
R_p/R_s	131 mΩ/117 mΩ				
R_{lp}/R_{ls}	210 mΩ/200 mΩ				
C_p/C_s	10 µH/7 µH				
f_{sw}	100 kHz				
$\phi_s, \delta_p, \delta_s$	0.13, 0.72, 0.52 rad				
C_{ip}	~0 pF	50 pF	200 pF	200 pF	200 pF
C_{is}	~0 pF	60 pF	240 pF	240 pF	240 pF
C_{ps}	~0 pF	1100 pF	1100 pF	100 pF	1000 pF
$i_{p,RMS}/i_{s,RMS}$	2.391 A/ 2.389 A	2.396 A/ 2.397 A	2.418 A/ 2.426 A	2.419 A/ 2.427 A	2.418 A/ 2.426 A
$P_{p,ac}/P_{s,ac}$	218 W/214 W	218 W/ 214 W	217.63 W/ 213.78 W	217.6 W/ 213.8 W	217.6 W/ 213.8 W

illustrates the current waveforms, revealing the effect of resonance between the intra-winding capacitances and port inductances. The high-frequency ringing in the primary inductor current i_p occurs at 4.6 MHz, corresponding to the resonance between C_{ip} and L_p. Similar observations are made for the secondary inductor current i_s. During test case-2, both C_{ip} and C_{is} are made four times while keeping C_{ps} constant at 1.1 nF. From the idea of resonance, now the frequency of ringing should be halved while the ringing amplitude will become higher. This, prediction is bolstered by the inductor current waveforms of Fig. 7.8a, where the i_p ringing frequency is observed as 2.3 MHz.

The effect of inter-winding capacitance variation on constructed DAB inductor currents is portrayed in Fig. 7.8b, corresponding to test case-4 and 5, where C_{ip} and C_{is} are kept at their nominal values of 100 and 70 pF, respectively, while C_{ps} was varied from 100 to 1000 pF. It is observed that the C_{ps} variation has minimal effect on the current shape determination.

From Table 7.2, it is observed that the inductor current RMS values and the input and output ac power of the DAB network are not significantly influenced by transformer parasitics due to the relatively low switching frequency of 100 kHz. At this frequency, the inductive impedance dominates the capacitive non-idealities for many harmonic orders.

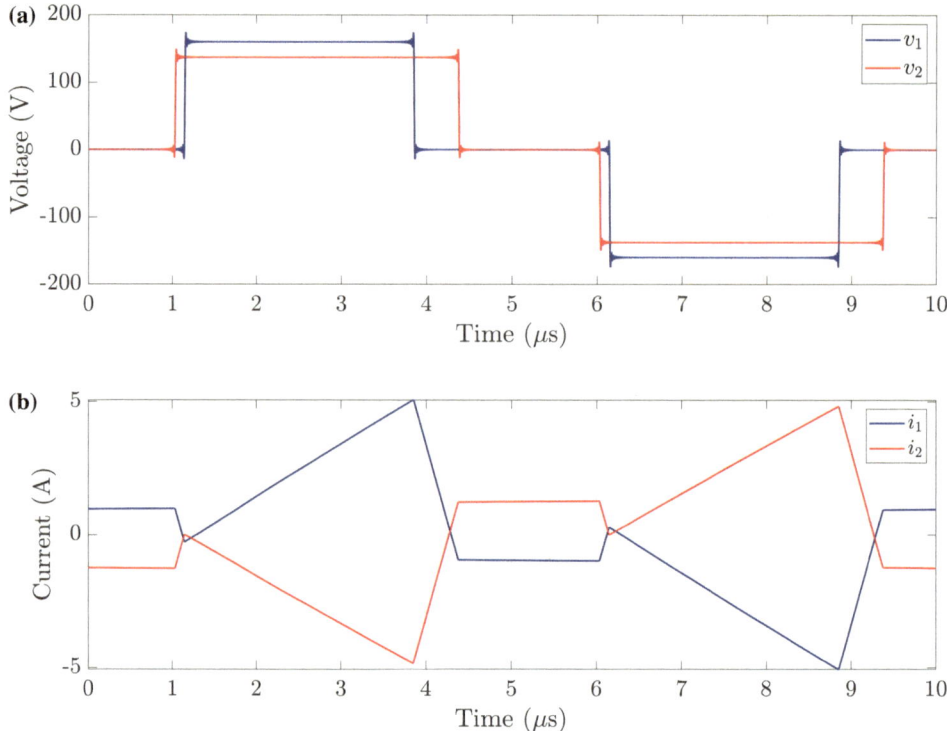

Fig. 7.7 **a** Equivalent 2-port generalized parasitic-inclusive MAB network. **b** Simplified T-network equivalent of 2-port generalized parasitic-inclusive MAB circuit (here, $i_1 = i_p$ and $i_2 = i_s$)

Thus, the power transfer and RMS currents can be estimated accurately using an ideal model for such a DAB converter. However, for accurately estimating switching losses and inferring the ZVS capabilities of the switching legs, it is essential to consider a model that includes non-ideal parasitic elements to predict the switching instant currents accurately.

The MATLAB script that is used to reconstruct the voltage and current waveforms of a 2-port MAB converter utilizing the transformer parasitic inclusive model is attached in the appendix section.

7.3   Inclusion of Deadtime Effects into Switching Voltage and Tank Current Reconstruction

Apart from transformer parasitics, the circuit dynamics during the finite deadtime between the gating signals of same switching leg switches shape the switching node voltages as well as the high-frequency currents. It is important to incorporate the deadtime effect into the modeling of bridge voltage waveforms especially when the deadtime to the switching

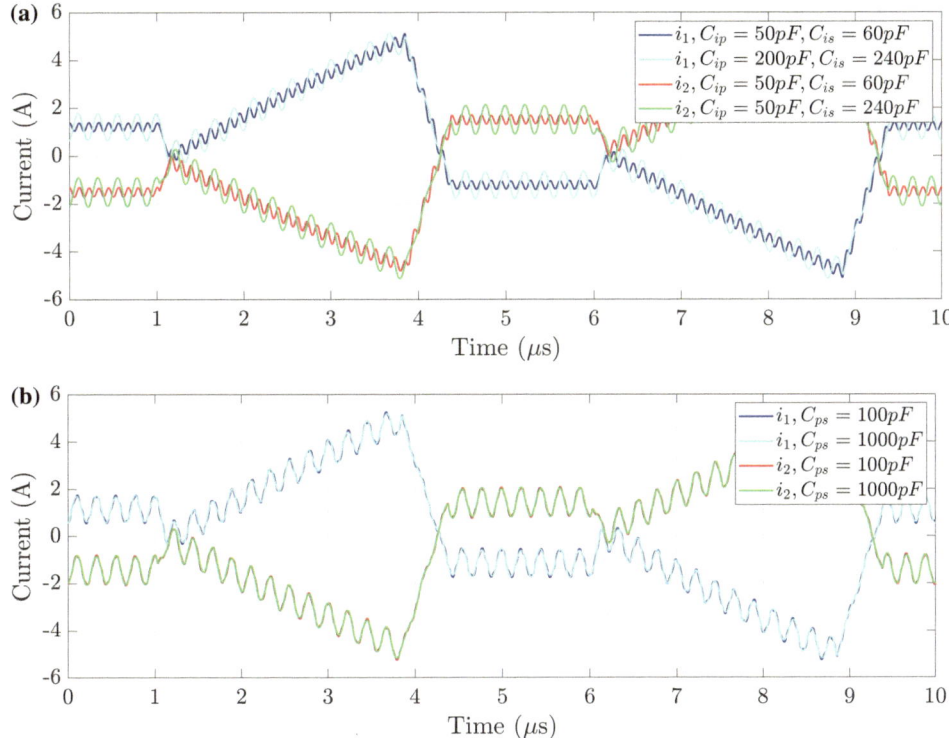

Fig. 7.8 **a** Equivalent 2-port generalized parasitic-inclusive MAB network (here, $i_1 = i_p$ and $i_2 = i_s$). **b** Simplified T-network equivalent of 2-port generalized parasitic-inclusive MAB circuit (here, $i_1 = i_p$ and $i_2 = i_s$)

period ratio is relatively higher in a high-frequency power conversion. Inaccurate bridge voltage reconstruction would lead to erroneous calculation of power flow in the converter as well as sub-optimal control variable synthesis that could lead to the loss of ZVS and/ or higher RMS currents.

For investigating the circuit dynamics during the dead time interval, here, we are considering a schematic of a two-port full-bridge based MAB converter, as shown in Fig. 7.9.

The circuit under consideration has an L-C tank on the primary side, a magnetizing inductance, and a power transfer inductor on the secondary side. For a DAB, the primary tank capacitance will be high enough to not cause any resonant frequency comparable with switching frequency and would be primarily serving DC blocking. Here, the dead-time period of T_d is assumed to occur at the rising edge of the gate pulses. In Fig. 7.10, an equivalent circuit is displayed that correlates to a mode transition where the primary bridge voltage v_p goes from $-V_{in}$ to V_{in} during the deadtime duration between turn-off

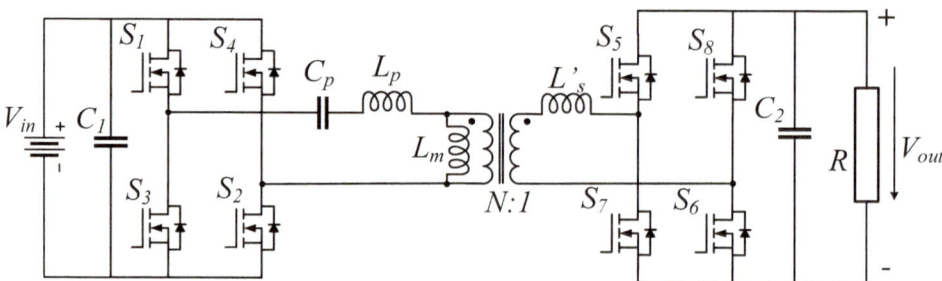

Fig. 7.9 Full-bridge based two port MAB converter

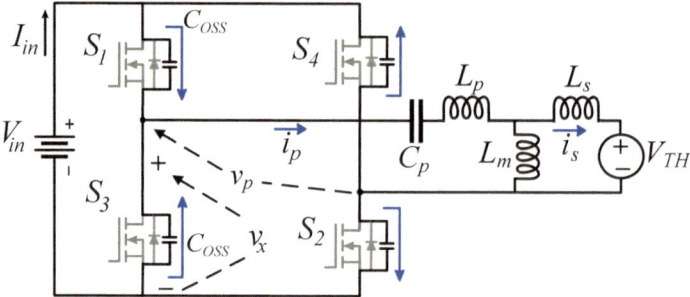

Fig. 7.10 Primary bridge equivalent circuit during deadtime interval between S_3 and S_4 turn-off and turn-on of S_1 and S_2

of S_3 and S_4 and turn-on of S_1 and S_2. This circuit analysis assumes that the tank current direction is favorable for ZVS transition of the S_1 and S_2, or $i_p < 0$.

During the deadtime T_d, the drain-source voltage of S_1 and S_2 is denoted by $(V_{in} - x)$, while the same for S_3 and S_4 is denoted by x that varies from 0 to V_{in}. Further, V_{TH} represents the secondary bridge voltage v_s during the switching instant reflected to primary side. The secondary port voltage at the primary switching transition would depend on the leading or lagging nature of the secondary bridge. If secondary lags primary, then V_{TH} will become $-NV_{out}$, where N is the turns ratio between the primary and secondary windings. Otherwise, a leading secondary will have V_{th} of $+NV_{out}$, when the primary bridges transitions from $-V_{in}$ to V_{in}. Considering $V_{TH} = -NV_{out}$, The following equations can be formulated based applying KVL on the equivalent circuit shown in Fig. 7.10.

$$V_p = 2x - V_{in} = L_p \frac{di_p}{dt} + \frac{1}{C_p} \int i_p dt + L_m \frac{di_p}{dt} - L_m \frac{di_s}{dt} \tag{7.31}$$

$$L_m \left(\frac{di_p}{dt} - \frac{di_s}{dt} \right) = L_s \frac{di_s}{dt} - NV_{out} \tag{7.32}$$

Sum of the currents responsible for S_1 body capacitor discharging and S_3 body capacitor charging is i_p, which gives the following relation.

$$-2C_{oss}\frac{dx}{dt} = i_p \tag{7.33}$$

In order to reconstruct the bridge voltage waveform, a closed-form expression of $x(t)$ needs to be found out using the above set of ordinary differential equations. Differentiating (7.31), yields:

$$2\frac{dx}{dt} = L_p\frac{d^2i_p}{dt^2} + \frac{i_p}{C_p} + L_m\frac{d^2i_p}{dt^2} - L_m\frac{d^2i_s}{dt^2} \tag{7.34}$$

Substituting $\frac{dx}{dt}$ in terms of i_p converts (7.34) into the expression below.

$$\left(L_p + L_m||L_s\right)\frac{d^2i_p}{dt^2} + \frac{i_p}{C_{eq}} = 0 \tag{7.35}$$

where, $C_{eq} = \frac{1}{\frac{1}{C_p} + \frac{1}{2C_{oss}}}$ i.e., the series equivalent of C_p and $2C_{oss}$.

General solution of (7.35) can be expressed in (7.36) as a sinusoid with frequency matching the resonant oscillation frequency of $L_{eq} = L_p + L_m||L_s$ and C_{eq}, i.e., $\omega_0 = \frac{1}{2\pi\sqrt{L_{eq}V_{eq}}}$

$$i_p(t) = A\sin(\omega_0 t) + B\cos(\omega_0 t) \tag{7.36}$$

At $t = 0$, the initial condition $i_p(0) = I_0$ can be calculated using a standard GHA model described in earlier chapters. That solves for the coefficient $B = I_0$.

Further, the drain-source voltage of S_3, $x(t)$ can be expressed as:

$$x(t) = \frac{-1}{2C_{oss}}\int i_p dt = \frac{1}{2C_{oss}}\left[\frac{A}{\omega_0}\cos(\omega_0 t) - \frac{B}{\omega_0}\sin(\omega_0 t) + K\right] \tag{7.37}$$

Applying the initial condition $x(0) = 0$, the integration constant K can be solved as: $K = -\frac{A}{\omega_0}$ that refines the solution in (7.37) as follows.

$$x(t) = \frac{1}{2C_{oss}}\left[\frac{A}{\omega_0}(\cos(\omega_0 t) - 1) - \frac{I_0}{\omega_0}\sin(\omega_0 t)\right] \tag{7.38}$$

At $t = 0$, (7.31) would turn into the initial condition of equivalent transition dynamics, formulated in (7.39), using the relation in (7.32) and $x(0) = 0$.

$$L_{eq}\frac{di_p}{dt}(0) + V_{c0} - \frac{L_m}{L_m + L_s}NV_{out} + V_{dc} = 0 \tag{7.39}$$

V_{c0} is the initial condition of the capacitor C_p voltage and can be calculated using GHA derived circuit modeling. Now, using (7.39) and (7.36), the coefficient A can be computed as: $A = \frac{L_{eq}}{L_S} N V_{out} - V_{in} - V_{c0}$. Thus, the final expression of $x(t)$ would turn out as follows:

$$x(t) = \frac{1}{2C_{oss}} \left[\frac{\frac{L_{eq}}{L_S} N V_{out} - V_{in} - V_{c0}}{\omega_0^2 L_{eq}} (\cos(\omega_0 t) - 1) - \frac{I_0}{\omega_0} \sin(\omega_0 t) \right]. \qquad (7.40)$$

Now, keeping the direction of the tank current i_p in mind the solution of v_p can be represented as:

$$\left\{ \max\left[V_{in}, \left\{ \frac{1}{C_{oss}} \left[\frac{\frac{L_{eq}}{L_S} N V_{out} - V_{in} - V_{c0}}{\omega_0^2 L_{eq}} (\cos(\omega_0 t) - 1) - \frac{I_0}{\omega_0} \sin(\omega_0 t) \right] - V_{in} \right\} \right], \right.$$

$$\text{if } i_p < 0; 0 < t < T_d V_{in},$$
$$\text{if } i_p > 0; 0 < t < T_d V_{in},$$
$$\text{if } t > T_d$$

$$(7.41)$$

As evident from (7.41), the transition segment of the bridge voltage would be a chopped part of the sinusoid with a specific resonant oscillation frequency. If $x(t)$ touches V_{in} i.e., $v_p(t)$ hits V_{in} before the deadtime ends, a ZVS turn-on for S_1 and S_2 is indicated, and v_p will take a value of $V_{in} + 2V_f$ due to forward conduction of the body diodes with voltage drop of V_f each. If $x(t)$ at the end of deadtime still falls short of V_{in}, V_p would not be able to hit V_{in} level as well, and a non-ZVS turn on with partial V-I overlap loss would occur for S_1 and S_2.

Likewise, similar modeling can be performed for other switching transition modes in the primary bridge. Moreover, same modeling approach will be taken when the secondary side full-bridge is switching while the primary port is maintaining a constant voltage of $+V_{in}$ or $-V_{in}$ depending on its lagging/leading nature.

The derived voltage waveform during deadtime before any MOSFET S_x turns on ($v_{p,dead,Sx}$), when added with the ideal quasi-square voltage waveform, yields the actual voltage waveform, as illustrated in Fig. 7.11.

Using the relations given in (7.41), $v_{p,dead,Sx}$ can also be expanded in Fourier series with its k-th harmonic component to be $\overrightarrow{V_{p,dead,Sx,k}}$. Now, the modified k-th harmonic bridge voltages will be:

$$\overrightarrow{V_{p,mod,k}} = \overrightarrow{V_{p,ideal,k}} + \overrightarrow{V_{p,dead,S1T,k}} + \overrightarrow{V_{p,dead,S2T,k}} + \overrightarrow{V_{p,dead,S1B,k}} + \overrightarrow{V_{p,dead,S2B,k}}; \qquad (7.42)$$

$$\overrightarrow{V_{s,mod,k}} = \overrightarrow{V_{s,ideal,k}} + \overrightarrow{V_{p,dead,S3T,k}} + \overrightarrow{V_{p,dead,S4T,k}} + \overrightarrow{V_{p,dead,S3B,k}} + \overrightarrow{V_{p,dead,S4B,k}}. \qquad (7.43)$$

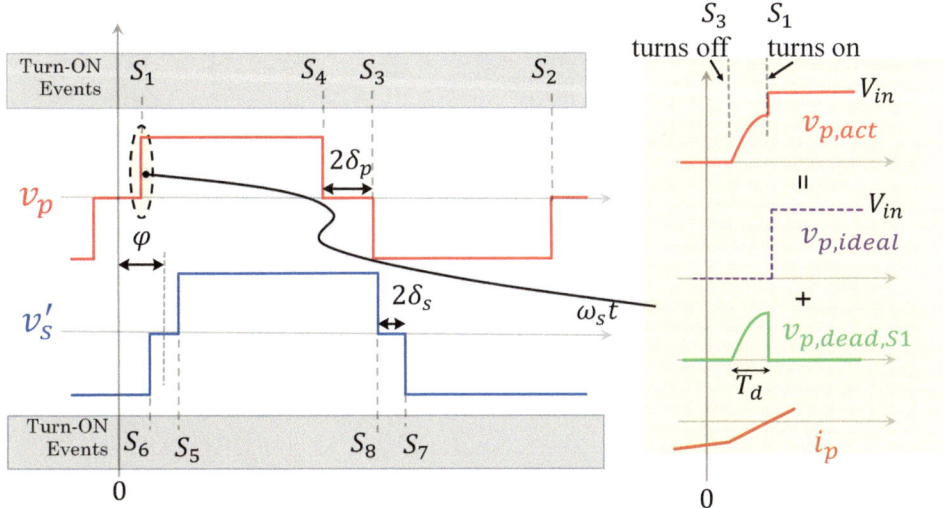

Fig. 7.11 Non-ideal switching transition waveform construction and definition of $v_{p,dead,S1T}$

Here, $\overrightarrow{V_{p,ideal,k}} = \frac{4V_{in}}{k\pi}\cos(k\delta_p + k\omega_s T_d/2)\angle 0^o$ and $\overrightarrow{V_{s,ideal,k}} = \frac{4V_{out}}{k\pi}\cos(k\delta_s + k\omega_s T_d/2)\angle - k\varphi$. Further, based on the modified bridge voltages the MAB tank currents and power flows are recalculated using the GHA model, described earlier.

The constructed HF DAB currents and bridge voltages are displayed in Fig. 7.12 showing a comparison between the ideal lossless model (ILM), the proposed NIFDM and the experimental results for a 160V-100V/200W power conversion. It is noticed that the NIFDM derived waveforms closely replicate the hardware waveforms for a partial and full ZVS scenario at the primary and secondary bridges, correspondingly. The error in estimating output power using the NIFDM at this operating condition is measured as 0.43% which is 16.2% lower compared to ILM estimated load power.

For a 100kHz isolated power conversion of 160V to 100V with a 7:5 transformer turns ratio at 200W with deadtime durations of 100ns on both primary and secondary sides, the bridge voltages accounting for the deadtime effects are shown in Fig. 7.12. It is noticed that the transformer parasitics and deadtime-effect inclusive modeled bridge voltages closely replicate the hardware waveforms for a partial and full ZVS scenario at the primary and secondary bridges, correspondingly, whereas the ideal quasi-square shaped voltages derived from the GHA based model fails to catch the deadtime effects that is crucial in deciding the soft-switching transitions as well as total power flow in the system.

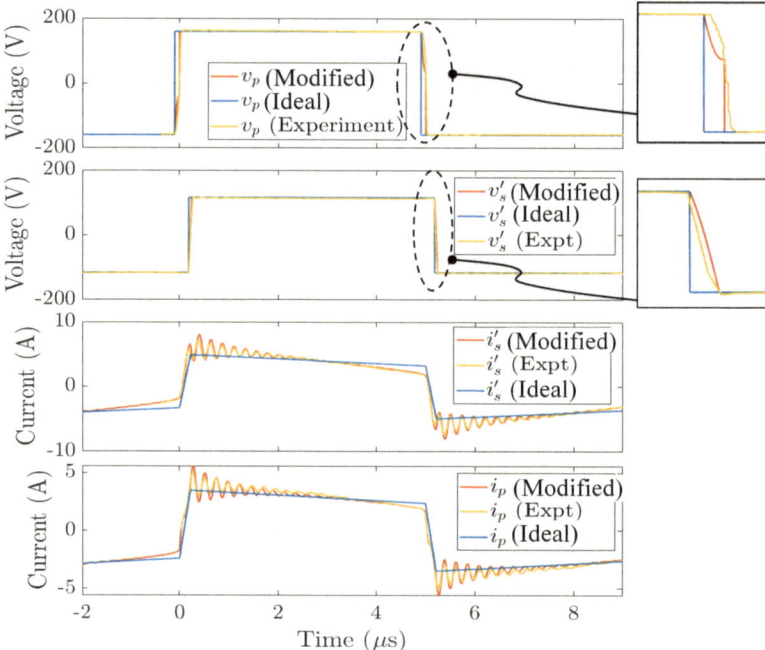

Fig. 7.12 Constructed DAB tank current and bridge voltage waveshapes including effects of transformer parasitics and deadtime in comparison to experimental waveforms

7.4 Loss Function Optimization of Capacitive Parasitics and Deadtime-Inclusive 2-Port MAB Converters

It is understandable from the previous analysis that for high-frequency converters, the transformer parasitics and a finite deadtime can become critical in determining the power flow among the converter ports, the port current and voltage shapes; thus, it also influences the losses in the system. Therefore, by meticulously reconstructing the AC port voltages and currents of the 2-port MAB and accounting for the transformer parasitics, the converter losses can be optimized using more precise loss models.

With the established two-port model given in Fig. 7.3a, total conduction loss in the switching network can be derived as a function of the control variables under any steady state operating condition.

$$F_{cond}\left(\delta_p, \delta_s, \varphi_s, f_{sw}\right) = i_{p,RMS}^2\left(R_p + R_{lp}\right) + i_{s,RMS}^2\left(R_s + R_{ls}\right); \qquad (7.44)$$

where R_p and R_s are the lumped equivalent resistances of the primary and secondary side, considering MOSFET on-state resistances and the dc/ac resistances of the passive elements. R_{lp} and R_{ls} are the ac resistances of the transformer primary and secondary windings, respectively. The $i_{p,RMS}$ and $i_{s,RMS}$ are computed using (7.27)–(7.28).

Additionally, total switching loss incurred in the MOSFETs due the overlap between the device current and voltage during the hard turn-on and hard turn-off switching instants for each HF switching leg is computed using (7.45).

$$
F_{sw,v-i}(\delta_p, \delta_s, \varphi_s, f_{sw}) = \sum_{x=1}^{2} [2V_{pdc}|i_p(\tau_{p,x})|f_{sw}t_{off,p,x} + 2V_{pdc}|i_p(\tau_{p,x})|f_{sw}t_{on,p,x} \cdot hard(p,x)]
$$
$$
+ [2V_{sdc}|i_s(\tau_{s,x})|f_{sw}t_{off,s,x} + 2V_{sdc}|i_s(\tau_{s,x})|f_{sw}t_{on,s,x} \cdot hard(p,x)].
$$
$$
(7.45)
$$

Here, $hard(p,x)$ and $hard(p,x)$ can be 0 or 1, depending on if the xth ($x = 1, 2$) switching leg of the primary and secondary H-bridge is experiencing fully zero-current or zero-voltage turn-on, or not. As for a FET turn-on, $hard(p,x) = 0$ will hold in case of (a) fully soft-switching with zero-voltage turn on or (b) when the instantaneous tank current polarity is favorable to discharge the FET body capacitor and the rising edge of gate-source appears when the falling drain-source voltage waveform reaches its valley i.e., when the FET body capacitor current has decreased to zero. The latter is a partial ZVS scenario with ZCS turn-on and hence zero V-I product, and the major dominant loss will be CV^2 loss, as expressed in (7.49). $hard(p,x) = 1$ will hold true when there is not sufficient dead-time for the FET body capacitor voltage to completely discharge and there is a non-zero source-to-drain current at the rising edge instant of gate-source voltage. If the tank current polarity is not favorable for ZVS turn-on, there will be full overlap between voltage fall and current rise, and $hard(p,x)$ will be 1. The conditions for hard turn-on of any switching leg devices in a L-tank DAB and LC-tank CLLC are identified and reported previous chapters. The turn-off current of switching leg-x of primary port is identified as $|i_{p,x}(\tau_{p,x})|$, where $\tau_{p,x}$ denotes the switching instants as given in Table 7.3. $|i_{p,x}(\tau_{p,x})|$ can be calculated using (7.23) Similarly, the turn-off currents of the secondary side legs are represented as $|i_{s,x}(\tau_{s,x})|$, $\tau_{s,x}$ denoting the switching instants. Further, the device turn-on and off times, $t_{on,i,x}$ and $t_{off,i,x}$ depend on the device parameters (C_{oss}, C_{iss}, C_{rss}) and used gate resistances ($R_{G,on}$ and $R_{G,off}$) by following the relations given in (7.46) and (7.47), respectively [8].

$$
t_{on,i,x} = R_{g,on,i}C_{iss,i}\ln\left(\frac{V_{dr,i} - V_{th,i}}{V_{dr,i} - V_{pl,i}}\right) + \frac{C_{rss,i}(V_i - R_{on,i}|i_{i,x}(\tau_{i,x,on})|)R_{g,on,i}}{V_{dr,i} - V_{pl,i}}; \quad (7.46)
$$

$$
t_{off,i,x} = R_{g,on,i}C_{iss,i}\ln\left(\frac{V_{dr,i} - V_{th,i}}{V_{dr,i} - V_{pl,i}}\right) + \frac{C_{rss,i}(V_i - R_{on,i}|i_{i,x}(\tau_{i,x,on})|)R_{g,on,i}}{V_{dr,i} - V_{pl,i}}. \quad (7.47)
$$

Here, $V_{dr,i}, V_{th,i}$ and $V_{pl,i}$ represent the gate drive voltage, gate threshold voltage and plateau voltage of the switching MOSFET connected at port-i of the MAB converter. Furthermore, the total number of switching legs undergoing soft turn-on in the converter

Table 7.3 Switching instants of the switching legs in a 2-port MAB

Switch	$\tau_{i,x,on}$	$\tau_{i,x,off}$
S_1	$\delta_p + T_d$	$\pi + \delta_p$
S_4	$\pi - \delta_p + T_d$	$-\delta_p$
S_5	$\varphi + \delta_s + T_d$	$\pi + \varphi + \delta_s$
S_8	$\varphi + \pi - \delta_s + T_d$	$\varphi - \delta_s$

can be determined as,

$$F_{ZVS}\left(\delta_p,\delta_s,\varphi_s,f_{sw}\right) = \sum_{x=1}^{2} ZVS(p,x) + ZVS(s,x) \tag{7.48}$$

Apart from the losses incurred by overlap between switching voltage and current during any switching event, the incomplete charge and discharge of the MOSFET's body capacitors (C_{oss}) also add on to the switching loss under and partial or full hard-switching event. This C_{oss} loss in the converter can be estimated as,

$$F_{Coss}\left(\delta_p,\delta_s,\varphi_s,f_{sw}\right) = \sum_{i=p,s} \sum_{x=1}^{2} 2\left[E_{oss,j}\left(\Delta V_j^x\right) + \left\{Q_{oss,j}(V_j) - Q_{oss,j}\left(V_j - \Delta V_j^x\right)\right\} \cdot V_j\right.$$

$$\left. - \left\{E_{oss,j}(V_j) - E_{oss,j}\left(V_j - \Delta V_j^x\right)\right\} + \frac{1}{2}C_{par,j}V_j^2\right] \cdot f_{sw} \tag{7.49}$$

Here, $E_{oss,j}\left(\Delta V_j^x\right)$ denotes the energy stored in the nonlinear capacitance C_{oss} of the MOSFET connected at x-th switching leg of j-side bridge ($j = p$ for primary bridge, $j = s$ for secondary bridge) due to the presence of ΔV_j^x voltage across it during its turn-on instant. This remaining drain-source voltage ΔV_j^k at turn-on instant can be found out by solving the energy balance as shown in Chap. 2 (ref (2.72)). For any drain-source voltage V_{DS}, the stored energy at C_{oss} is $E_{oss}(V_{DS}) = \int_0^{V_{DS}} C_{oss}(v) \cdot v dv = \frac{1}{2}C_E(V_{DS}) \cdot V_{DS}^2$ and can be calculated from the $C_{oss} - V_{DS}$ plot given in the device datasheet. Similarly, $Q_{oss}(V_{DS})$ denotes the amount of charge stored in C_{oss} due to V_{DS} voltage present across it and can be calculated as, $Q_{oss}(V_{DS}) = \int_0^{V_{DS}} C_{oss}(v) \cdot dv = C_Q(V_{DS}) \cdot V_{DS}$. The $C_{par,j}$ is the parasitic switching node capacitance of j-bridge that appears due to the printed circuit board design, parasitic capacitances of the transformer etc. and needs to be charged and discharged everytime during any switching commutation. In case of complete soft-switching transition, ΔV_j^x will be 0, while, for a complete hard-switching, $\Delta V_j^x \rightarrow V_j$. The conditions for complete ZVS in such two-port MAB are outlined in Chap. 2.

Thus, the total switching loss in the converter can be expressed as,

$$F_{sw}\left(\delta_p,\delta_s,\varphi_s,f_{sw}\right) = F_{Coss}\left(\delta_p,\delta_s,\varphi_s,f_{sw}\right) + F_{sw,v-i}\left(\delta_p,\delta_s,\varphi_s,f_{sw}\right). \tag{7.50}$$

It is understood that the losses incurred by the switching network heavily depend on the control parameters. For a two-port MAB, there are four control variables that can be adaptively adjusted to meet load demands while modifying the bridge voltage and current waveforms to address different converter loss scenarios. By identifying this scope of optimizing the converter's control parameters, i.e., the PWM phase-duty and frequency, for improved system performance—reducing conduction and switching losses while increasing soft-switching events, a control optimization program can be developed. It optimizes control variables $x = [\delta_p, \delta_s, \varphi_s, f_{sw}]$ for preferred objective functions $F_{cond}(x)$ or $F_{sw}(x)$ or $F_{ZVS}(x)$ or $F_{tot}(x)$, i.e., $F_{cond}(x) + F_{sw}(x)$—corresponding to conduction loss, switching loss, ZVS leg count, total switching network loss, respectively, derived using (7.44)–(7.50).

The output power flow constraint at the two-port MAB dc port can be computed using (7.38), when the secondary port works as an output.

$$P_{s,out} = P_{s,ac} - \sum_{x=1}^{2} \left[2\left[E_{oss,s}\left(\Delta V_s^k\right) + \left\{Q_{oss,s}(V_{sdc}) - Q_{oss,s}\left(V_{sdc} - \Delta V_s^k\right)\right\} \cdot V_{sdc} \right.$$

$$- \left\{E_{oss,s}(V_{sdc}) - E_{oss,s}\left(V_{sdc} - \Delta V_s^k\right)\right\} + \frac{1}{2}C_{par,s}V_{sdc}^2 \right] \cdot f_{sw}$$

$$\left. + \left\{2V_{sdc}\left|i_s(\tau_{s,x})\right|f_{sw}t_{off,s,x} + 2V_{sdc}\left|i_s(\tau_{s,x})\right|f_{sw}t_{on,s,x} \cdot hard(p,x)\right\}\right] \tag{7.51}$$

Here, $P_{s,out}$ is the load connected at the secondary dc output; $P_{s,ac}$ is the average ac power fed to the secondary bridge, computed from (7.30); the switching and C_{oss} related losses in the secondary side H-bridge are calculated as $\sum_{x=1}^{2}\left[2V_{sdc}\left|i_s(\tau_{s,x})\right|f_{sw}t_{off,s,x} + 2V_{sdc}\left|i_s(\tau_{s,x})\right|f_{sw}t_{on,s,x} \cdot hard(p,x)\right]$ and $\sum_{x=1}^{2}\left[2[E_{oss,s}(\Delta V_s^k) + \{Q_{oss,s}(V_{sdc}) - Q_{oss,s}(V_{sdc} - \Delta V_s^k)\} \cdot V_{sdc} - \{E_{oss,s}(V_{sdc}) - E_{oss,s}(V_{sdc} - \Delta V_s^k)\} + \frac{1}{2}C_{par,s}V_{sdc}^2\right] \cdot f_{sw}]$, respectively.

Moreover, the upper and lower bounds x are keyed into the program by the user according to the design demand. Now solving this single objective single constrained multi-variate optimization problem, we can attain the optimal control variable set that leads to the minimal loss point operation of the converter.

7.5 Conclusions

This chapter presented a modeling technique for 2-port resonant and non-resonant MABs, such as DAB, CLLC, and LLC converters, that accounts for transformer parasitics, specifically inter- and intra-winding capacitances. The study revealed that in transformers with a planar structure, where parasitic capacitances between windings are significantly high, the port current waveforms can become distorted due to high-frequency resonance between inductive elements and these parasitic capacitances. This distortion, or ringing, can generate both common mode and differential mode noise in the converter. Additionally, the non-ideality-inclusive current waveforms during switching transitions may result in a loss of soft-switching, an issue that could go unnoticed if the converter is modeled without considering capacitive parasitics of the transformer. Furthermore, this chapter outlines an approach for minimizing converter losses by optimally and precisely selecting phase-shift and frequency control variables, utilizing a more realistic and non-ideal model of the converter.

References

1. J. Biela and J. W. Kolar, "Using transformer parasitics for resonant converters - a review of the calculation of the stray capacitance of transformers," *Fortieth IAS Annual Meeting. Conference Record of the 2005 Industry Applications Conference, 2005.*, Hong Kong, China, 2005, pp. 1868–1875 Vol. 3, https://doi.org/10.1109/IAS.2005.1518701.
2. J. G. Hayes, N. O'Donovan and M. G. Egan, "The extended T model of the multiwinding transformer," *2004 IEEE 35th Annual Power Electronics Specialists Conference (IEEE Cat. No.04CH37551)*, Aachen, Germany, 2004, pp. 1812–1817 Vol.3, https://doi.org/10.1109/PESC.2004.1355391.
3. Z. Ouyang, O. C. Thomsen and M. A. E. Andersen, "Optimal Design and Tradeoff Analysis of Planar Transformer in High-Power DC–DC Converters," in *IEEE Transactions on Industrial Electronics*, vol. 59, no. 7, pp. 2800-2810, July 2012, https://doi.org/10.1109/TIE.2010.2046005.
4. A. Chandwani and A. Mallik, "Fabrication Tradeoff Based Optimal Synthesis of Winding Configurations for Planar Transformer in CLLC DC–DC Converter," in *IEEE Transactions on Power Electronics*, vol. 38, no. 5, pp. 6243-6258, May 2023, https://doi.org/10.1109/TPEL.2023.3241135.
5. O. Zayed, A. Elezab and M. Narimani, "A Co-Planar Transformer with Ultra-Low Parasitic Capacitance for EV Chargers," in *IEEE Transactions on Transportation Electrification*, https://doi.org/10.1109/TTE.2024.3384962.
6. H. Zhang and S. Wang, "Two-capacitor transformer winding capacitance models for common-mode EMI noise analysis in isolated DC-DC converters," *2016 IEEE Energy Conversion Congress and Exposition (ECCE)*, Milwaukee, WI, USA, 2016, pp. 1-8, https://doi.org/10.1109/ECCE.2016.7855538.
7. M. A. Saket, M. Ordonez, M. Craciun and C. Botting, "Improving Planar Transformers for LLC Resonant Converters: Paired Layers Interleaving," in *IEEE Transactions on Power Electronics*, vol. 34, no. 12, pp. 11813-11832, Dec. 2019, https://doi.org/10.1109/TPEL.2019.2903168.
8. A. Mallik, Lecture Notes, Power Electronics EGR 494/598, Switching Loss Modeling and Calculation. https://peacelabasu.s3-us-west-1.amazonaws.com/Courses/EGR+494_598/Lecture-2+Loss+calculation.pdf.

Appendix

Chapter 3 Example Codes

MATLAB Script to Construct Voltage and Current Waveforms for a DAB Converter Under Any Operating Condition

```
% MATLAB Script to Generate Voltage and Current Waveforms for a DAB Converter
% Author: Saikat Dey
% Arizona State University
% This script calculates and plots the switching voltages and current for
% a Dual Active Bridge (DAB) converter.

function DAB_vol_curr_plot(Vp, Vs, n, Lp, Ls, delta1, delta2, phi, fsw)

% Input Parameters:
% Vp     - Primary voltage
% Vs     - Secondary voltage
% n      - Transformer turns ratio (n = Np/Ns)
% Lp     - Primary inductance (uH)
% Ls     - Secondary inductance (uH)
% delta1 - Duty cycle for primary side
% delta2 - Duty cycle for secondary side
% phi    - Phase shift between primary and secondary
% fsw    - Switching frequency (kHz)
```

© The Editor(s) (if applicable) and The Author(s), under exclusive license
to Springer Nature Switzerland AG 2025
A. Mallik and S. Dey, *Switching Modulator Optimization in Isolated Power Converters*,
Synthesis Lectures on Power Electronics, https://doi.org/10.1007/978-3-031-81576-8

```matlab
% Derived Parameters
V1 = Vp;                               % Set primary voltage
G = Vs / (n * V1);                     % Gain factor
L1 = Lp * 1e-6;                        % Convert primary inductance to Henries
L2 = Ls * ((1/n)^2) * 1e-6;           % Convert secondary inductance to
Henries
L = L1 + L2;                           % Total inductance

% Time Vector and Angular Frequency
wg = 2 * pi * 1000 * fsw;              % Angular frequency in rad/s
theta = 0:(pi/(1000*fsw)):2*pi;        % Angle range for one period
t = theta / wg;                        % Time vector

% Initialize Summation Variables
sum12 = 0;
sumx = 0;
sumy = 0;

% Harmonic Summation Calculations (all done offline)
for k = 1:2:2001
    sumx = sumx + (4 * V1 / (pi * k)) * (cos(k * delta1) * sin(k * theta));
    sumy = sumy + (4 * V1 * G / (pi * k)) * (cos(k * delta2) * sin(k * (theta -
phi)));
    sum12 = sum12 + (4 * V1 / (pi * wg * k^2 * L)) * (-cos(k * delta1) * cos(k *
theta) + G * cos(k * delta2) * cos(k * (theta - phi)));
end

% Plotting
figure(1);

% Plot Primary and Scaled Secondary Switching Voltages
yyaxis left
plot(t, sumx, t, sumy * n, 'LineWidth', 2.5);
grid on;
ylabel('Switching Voltages (V)', 'Interpreter', 'latex', 'FontSize', 28);

% Plot Primary Side Current
yyaxis right
plot(t, sum12, 'LineWidth', 2.5);
grid on;
legend({'$v_{1}$','$v^{,}_{2}$','$i_{1}$'}, 'Location', 'northeast',
'Interpreter', 'latex');
ylabel('$i_{1}$ (A)', 'Interpreter', 'latex', 'FontSize', 28);

% Set Common Plot Features
ax = gca;
ax.TickLabelInterpreter = 'latex';
set(gca, 'FontSize', 24);
xlabel('$time (t)$ [sec]', 'Interpreter', 'latex', 'FontSize', 28);

end
```

MATLAB Script to Plot the RMS Optimized DAB Inductor Current and Modulation Variables for Varying Load and Gain Conditions

```
% Main Script for Optimizing and Plotting DAB Converter Modulation
TPS = [];
Po = 50;  % Desired output power in Watts

%% TPS %%
for i = 0.5:0.01:2
    G = i;
    del_1_range = pi/2;
    del_2_range = pi/2;
    % Perform optimization for TPS modulation
    TPS = [TPS; G, Po, optimize_rms(G, Po, del_1_range, del_2_range)];
end

%% DPS1 (delta2=0) %%
DPS1 = [];
for i = 0.5:0.01:2
    G = i;
    del_1_range = pi/2;
    del_2_range = 0;
    % Perform optimization for DPS1 modulation (delta2=0)
    DPS1 = [DPS1; G, Po, optimize_rms(G, Po, del_1_range, del_2_range)];
end
final1 = [final, final1];

%% DPS2 (delta1=0) %%
DPS2 = [];
for i = 0.5:0.01:2
    G = i;
    del_1_range = 0;
    del_2_range = pi/2;
    % Perform optimization for DPS2 modulation (delta1=0)
    row = [];
    DPS2 = [DPS2; G, Po, optimize_rms(G, Po, del_1_range, del_2_range)];
end
final2 = [final1, final2];

%% SPS %%
SPS = [];
for i = 0.5:0.01:2
    G = i;
    del_1_range = 0;
    del_2_range = 0;
    % Perform optimization for SPS modulation
    SPS = [SPS; G, Po, optimize_rms(G, Po, del_1_range, del_2_range)];
end

% Display final results
disp(SPS);

% Save results to CSV file
%csvwrite('DAB_Modulation_50W.csv', final3);

%% Plotting Figures %%
% Plotting Cost vs Gain (G)
figure(1);
plot(final3(:, 1), final3(:, 6), '-r', 'LineWidth', 2);
```

```
title('Irms vs G');
xlabel('Gain (G)');
ylabel('Cost Func, iLRMS (A)');
hold on;
grid on;
plot(TPS(:, 1), TPS(:, 6), '-r');
plot(DPS1(:, 1), DPS1(:, 6), '--b');
plot(DPS2(:, 1), DPS2(:, 6), '-.c');
plot(SPS(:, 1), SPS(:, 6), ':mo');
legend('TPS', 'DPS-1', 'DPS-2', 'SPS');

% Plotting delta1 vs Gain (G)
figure(2);
title('delta1 vs G');
xlabel('Gain (G)');
ylabel('delta1');
hold on;
grid on;
plot(TPS(:, 1), TPS(:, 3), '-r');
plot(DPS1(:, 1), DPS1(:, 3), '--b');
plot(DPS2(:, 1), DPS2(:, 3), '-.c');
plot(SPS(:, 1), SPS(:, 3), ':mo');
legend('TPS', 'DPS-1', 'DPS-2', 'SPS');

% Plotting delta2 vs Gain (G)
figure(3);
title('delta2 vs G');
xlabel('Gain (G)');
ylabel('delta2');
hold on;
grid on;
plot(TPS(:, 1), TPS(:, 4), '-r');
plot(DPS1(:, 1), DPS1(:, 4), '--b');
plot(DPS2(:, 1), DPS2(:, 4), '-.c');
plot(SPS(:, 1), SPS(:, 4), ':mo');
legend('TPS', 'DPS-1', 'DPS-2', 'SPS');

% Plotting phi vs Gain (G)
figure(4);
title('phi vs G');
xlabel('Gain (G)');
ylabel('phi');
hold on;
grid on;
plot(TPS(:, 1), TPS(:, 5), '-r');
plot(DPS1(:, 1), DPS1(:, 5), '--b');
plot(DPS2(:, 1), DPS2(:, 5), '-.c');
plot(SPS(:, 1), SPS(:, 5), ':mo');
legend('TPS', 'DPS-1', 'DPS-2', 'SPS');

% Optimization function to minimize RMS current
function [y] = optimize_rms(G, Po, del_1_range, del_2_range)
    % Setting up the objective function to minimize RMS current
    obj = @(x) rms_objective(x(1), x(2), x(3), G);

    % Optimization setup
```

```
    A = []; b = []; Aeq = []; beq = [];
    lb = [0, 0, -pi/2];                   % Lower bounds
    ub = [del_1_range, del_2_range, pi/2]; % Upper bounds
    nonlcon = @(x) power_constraint(x, Po, G);
    x0 = [0, 0, 0]; % Starting point for optimization

    % Perform the constrained optimization using fmincon
    x = fmincon(obj, x0, A, b, Aeq, beq, lb, ub, nonlcon);

    % Return the optimized variables and the objective function value
    y = [x(1), x(2), x(3), obj(x)];
end

% RMS current objective function
function y = rms_objective(delta1, delta2, phi, G)
    % Constants
    V1 = 160;              % Primary voltage in Volts
    fs = 1e5;              % Switching frequency in Hz
    L1 = 16e-6;            % Primary inductance in Henrys
    Ls = 15e-6;            % Secondary inductance in Henrys
    L2 = Ls * (7/5)^2;     % Secondary inductance adjusted for turns ratio
    L = L1 + L2;           % Total inductance in Henrys
    wg = 2 * pi * fs;      % Angular frequency

    % Initialize summation variable for harmonic series
    sum1 = 0;

    % Harmonic series calculation for RMS current
    for k = 1:2:100
        sum1 = sum1 + ((G * cos(k * delta2))^2 + (cos(k * delta1))^2 - ...
            2 * G * cos(k * delta1) * cos(k * delta2) * cos(k * phi)) * V1^2 /
k^4;
    end

    % Calculate RMS current
    i1rms = sqrt(sum1 * 8 / (pi^2 * wg^2 * L^2));
    i2rms = sqrt(sum1 * 8 / (pi^2 * wg^2 * L^2));

    % y = i1rms^2 + i2rms^2; % Objective: minimize the sum of squares of i1rms and
i2rms
    y = i1rms; % Objective: minimize the i1rms
end

% Power constraint function for the optimization
function [f, feq] = power_constraint(x, Po, G)
    % Extract optimization variables
    delta1 = x(1);
    delta2 = x(2);
    phi = x(3);

    % Constants
    V1 = 160;              % Primary voltage in Volts
    fs = 1e5;              % Switching frequency in Hz
    L1 = 16e-6;            % Primary inductance in Henrys
    Ls = 15e-6;            % Secondary inductance in Henrys
    L2 = Ls * (7/5)^2;     % Secondary inductance adjusted for turns ratio
```

```
        L = L1 + L2;          % Total inductance in Henrys

        % Initialize summation variable
        sum1 = 0;

        % Harmonic series calculation for power
        for k = 1:2:100
            sum1 = sum1 + (4 * G * V1^2 / (pi^3 * fs * L)) * ...
                (cos(k * delta1) * cos(k * delta2) * sin(k * phi)) / k^3;
        end

        % Set the power equality constraint
        feq(1) = sum1 - Po;

        % No inequality constraints
        f = [];
    end
```

MATLAB Script to Plot the RMS Optimized DAB Inductor Current and Modulation Variables for Varying Load and Gain Conditions in 3D for Different Modulation Techniques

```matlab
% Initialize parameters
G_range = 0.5:0.1:2;        % Gain range
Po_range = 50:50:400;       % Output power range (W)
num_G = length(G_range);
num_Po = length(Po_range);

% Preallocate matrices for storing results
phi_matrix = zeros(num_Po, num_G, 4);
delta1_matrix = zeros(num_Po, num_G, 4);
delta2_matrix = zeros(num_Po, num_G, 4);
irms_matrix = zeros(num_Po, num_G, 4);

% Modulation schemes
mod_schemes = {'TPS', 'DPS-1', 'DPS-2', 'SPS'};

% Loop through each modulation scheme
for scheme = 1:4
    for i = 1:num_G
        for j = 1:num_Po
            G = G_range(i);
            Po = Po_range(j);

            % Define delta1 and delta2 ranges based on the scheme
            switch scheme
                case 1   % TPS
                    del_1_range = pi/2;
                    del_2_range = pi/2;
                case 2   % DPS-1 (delta2 = 0)
                    del_1_range = pi/2;
                    del_2_range = 0;
                case 3   % DPS-2 (delta1 = 0)
                    del_1_range = 0;
                    del_2_range = pi/2;
                case 4   % SPS (delta1 = 0, delta2 = 0)
                    del_1_range = 0;
                    del_2_range = 0;
            end

            % Call optimize_rms function to get the optimal values
            results = optimize_rms(G, Po, del_1_range, del_2`range);

            % Store the results in the corresponding matrices
            delta1_matrix(j, i, scheme) = results(1);
            delta2_matrix(j, i, scheme) = results(2);
            phi_matrix(j, i, scheme) = results(3);
```

```
                irms_matrix(j, i, scheme) = sqrt(results(4));
            end
        end
    end

    % Figure 1: phi vs Output Power, Po vs Gain, G
    figure;
    for scheme = 1:4
        subplot(2,2,scheme);
        surf(G_range, Po_range, phi_matrix(:, :, scheme));
        title(['\phi vs Po vs G - ' mod_schemes{scheme}]);
        xlabel('Gain (G)');
        ylabel('Output Power (Po)');
        zlabel('\phi');
        grid on;
    end
    sgtitle('\phi vs Po vs G for Different Modulation Schemes');

    % Figure 2: delta1 vs Output Power, Po vs Gain, G
    figure;
    for scheme = 1:4
        subplot(2,2,scheme);
        surf(G_range, Po_range, delta1_matrix(:, :, scheme));
        title(['\delta_1 vs Po vs G - ' mod_schemes{scheme}]);
        xlabel('Gain (G)');
        ylabel('Output Power (Po)');
        zlabel('\delta_1');
        grid on;
    end
    sgtitle('\delta_1 vs Po vs G for Different Modulation Schemes');

    % Figure 3: delta2 vs Output Power, Po vs Gain, G
    figure;
    for scheme = 1:4
        subplot(2,2,scheme);
        surf(G_range, Po_range, delta2_matrix(:, :, scheme));
        title(['\delta_2 vs Po vs G - ' mod_schemes{scheme}]);
        xlabel('Gain (G)');
        ylabel('Output Power (Po)');
        zlabel('\delta_2');
        grid on;
    end
    sgtitle('\delta_2 vs Po vs G for Different Modulation Schemes');
```

```matlab
% Figure 4: Optimized iLRMS vs Output Power, Po vs Gain, G
figure;
for scheme = 1:4
    subplot(2,2,scheme);
    surf(G_range, Po_range, irms_matrix(:, :, scheme));
    title(['i_{LRMS} vs Po vs G - ' mod_schemes{scheme}]);
    xlabel('Gain (G)');
    ylabel('Output Power (Po)');
    zlabel('i_{LRMS} (A)');
    grid on;
end
sgtitle('i_{LRMS} vs Po vs G for Different Modulation Schemes');
```

Chapter 7 (Example Code)

Chapter 7 (example code):

```
% Optimized MATLAB Script for 2-Port MAB Converter Voltage and Current Modeling
% Author: Saikat Dey
% Arizona State University
% This script models voltage and current behavior in a 2-Port Multi-Active Bridge
(MAB) converter,
% considering parasitics.

% Input parameters:
% Vin: Input voltage (V)
% Vout: Output voltage (V)
% phi: Phase shift between the two bridges (rad)
% delta1: Duty cycle of the primary side (rad)
% delta2: Duty cycle of the secondary side (rad)
% fsw: switching frequency in kHz

function [y] = MAB_2_port_model_parasitics(Vin, Vout, phi, delta1, delta2, fsw)

% Parameters and constants
V1 = Vin;
n1 = 7;
n2 = 5;
V2 = n1/n2 * Vout;
fsw = fsw*1e3;                               % Switching frequency in Hz
w = 2 * pi * fsw;                            % Angular frequency in rad/s
t = linspace(0, 20e-6, 100000);             % Time vector from 0 to 20
microseconds
h = 101;                                     % no of harmonic considered

% Define Parasitic Resistances
Ron1 = 15*1e-3; Ron2 = (n1/n2)^2 * 4*1e-3;   % On-state Device Resistances
R_L1 = 500e-3; R_L2 = (n1/n2)^2 * 10e-3;     % Series Inductor dc-R
Rlp = 100e-3; Rls = (n1/n2)^2 * 70e-3;       % Transformer winding dc-R
esr_C1 = 5e-3; esr_C2 = (n1/n2)^2 *5e-3;     % dc-blocking/resonant Cap ESR
Rl1 = R_L1 + Rlp; Rl2 = R_L2 + Rls;

% Define Capacitances
C1 = (10) * 1e-6; C2 = (12) * 1e-6 / ((n1/n2)^2);  % Resonant or dc-blocking Caps
Cis = (100) * 1e-12; Cip = (100) * 1e-12; Cps = (1100) * 1e-12;        %
Transformer Parasitics

% Define Inductances (refered to primary)
L1 = (5.96) * 1e-6; L2 = (n1/n2)^2 * 2.85 * 1e-6;  % Series Port Inductances
Lm = 776.89e-6;                              % Magnetizing Inductance
Llp = (0.24) * 1e-6; Lls = (n1/n2)^2 * 0.12 * 1e-6; % Leakage Inductances

% Preallocate variables
Z12 = zeros(1, h); Z13 = zeros(1, h); Z23 = zeros(1, h);
Ai1 = zeros(1, h); Bi1 = zeros(1, h);
Ai2 = zeros(1, h); Bi2 = zeros(1, h);
i1_t = zeros(length(t), h); i2_t = zeros(length(t), h);
v1_t = zeros(length(t), h); v2_t = zeros(length(t), h);
sum_Pac_1 = 0; sum_Pac_2 = 0; sum_i1_RMS_squared = 0; sum_i2_RMS_squared = 0;

% Compute impedances and perform calculations
for k_idx = 1:h
    k = 2*k_idx-1;
```

```
ac = k/2; % ac-resistance factor, change as needed

% Impedance calculations for primary, secondary, and magnetizing branches of
Transformer
Zlp    = Rlp + 1i * w * k * Llp;
Zls    = Rls + 1i * w * k * Lls;
Zlm    = 1i * w * k * Lm;

% Parasitic capacitance impedance calculations
Z_Cip = 1 / (1i * w * k * Cip);
Z_Cis = 1 / (1i * w * k * Cis);
Z_Cps = 1 / (1i * w * k * Cps);

% star to delta
Z_L1 = Zlp + Zlm + (Zlp * Zlm) / Zls;
Z_L2 = Zls + Zlm + (Zls * Zlm) / Zlp;
Z_L3 = Zlp + Zls + (Zlp * Zls) / Zlm;

% Impedance calculations for the combined branches
Z_1  = (Z_L1 * Z_Cip) / (Z_L1 + Z_Cip);
Z_2  = (Z_L2 * Z_Cis) / (Z_L2 + Z_Cis);
Z_3  = (Z_L3 * Z_Cps) / (Z_L3 + Z_Cps);

Zp = 2 * Ron1 + Rl1*ac + esr_C1 + 1i * w * k * L1 + 1/(1i * w * k * C1);
Zs = 2 * Ron2 + Rl2*ac + esr_C2 + 1i * w * k * L2 + 1/(1i * w * k * C2);

Z_1_dash = Zp + Z_1*Z_3 / (Z_1+Z_2+Z_3);
Z_2_dash = Zs + Z_2*Z_3 / (Z_1+Z_2+Z_3);
Z_3_dash = Z_1*Z_2 / (Z_1+Z_2+Z_3);

Z_tot = Z_1_dash * Z_2_dash + Z_1_dash * Z_3_dash + Z_2_dash * Z_3_dash;

Z12(k_idx) = Z_tot / Z_3_dash;
Z23(k_idx) = Z_tot / Z_1_dash;
Z13(k_idx) = Z_tot / Z_2_dash;

v1_abs = 4 * V1 * cos(k * delta1) / (k * pi);
v2_abs = 4 * V2 * cos(k * delta2) / (k * pi);

Ai1(k_idx) = (v1_abs / abs(Z12(k_idx))) * cos(angle(Z12(k_idx))) - ...
             (v2_abs / abs(Z12(k_idx))) * cos(k * phi + angle(Z12(k_idx))) +
...
             (v1_abs / abs(Z13(k_idx))) * cos(angle(Z13(k_idx)));
Bi1(k_idx) = -(v1_abs / abs(Z12(k_idx))) * sin(angle(Z12(k_idx))) + ...
             (v2_abs / abs(Z12(k_idx))) * sin(k * phi + angle(Z12(k_idx))) -
...
             (v1_abs / abs(Z13(k_idx))) * sin(angle(Z13(k_idx)));
Ai2(k_idx) = -(v1_abs / abs(Z12(k_idx))) * cos(angle(Z12(k_idx))) + ...
             (v2_abs / abs(Z12(k_idx))) * cos(k * phi + angle(Z12(k_idx))) +
...
             (v2_abs / abs(Z23(k_idx))) * cos(k * phi + angle(Z23(k_idx)));
Bi2(k_idx) = +(v1_abs / abs(Z12(k_idx))) * sin(angle(Z12(k_idx))) - ...
             (v2_abs / abs(Z12(k_idx))) * sin(k * phi + angle(Z12(k_idx))) -
...
             (v2_abs / abs(Z23(k_idx))) * sin(k * phi + angle(Z23(k_idx)));
```

```matlab
    % Summing up the harmonics
    i1_t(:, k_idx) = Ai1(k_idx) * sin(k * w * t) + Bi1(k_idx) * cos(k * w * t);
    i2_t(:, k_idx) = Ai2(k_idx) * sin(k * w * t) + Bi2(k_idx) * cos(k * w * t);
    v1_t(:, k_idx) = v1_abs * sin(k * w * t);
    v2_t(:, k_idx) = v2_abs * sin(k * (w * t - phi));

    sum_i1_RMS_squared = sum_i1_RMS_squared + (Ai1(k_idx)^2 + Bi1(k_idx)^2);
    sum_i2_RMS_squared = sum_i2_RMS_squared + (Ai2(k_idx)^2 + Bi2(k_idx)^2);
    sum_Pac_1 = sum_Pac_1 + (v1_abs / sqrt(2)) * sqrt((Ai1(k_idx)^2 +
Bi1(k_idx)^2) / 2) * cos(atan2(Bi1(k_idx), Ai1(k_idx)));
    sum_Pac_2 = sum_Pac_2 + (v2_abs / sqrt(2)) * sqrt((Ai2(k_idx)^2 +
Bi2(k_idx)^2) / 2) * cos(k * phi + atan2(Bi2(k_idx), Ai2(k_idx)));
    end

  % RMS currents and power calculations
  I1_RMS = sqrt(sum_i1_RMS_squared / 2);
  I2_RMS = sqrt(sum_i2_RMS_squared / 2);
  Pac_1 = sum_Pac_1;
  Pac_2 = sum_Pac_2;

  % Final waveform sums
  i1_total_t = sum(i1_t, 2); i2_total_t = sum(i2_t, 2);
  v1_total_t = sum(v1_t, 2); v2_total_t = sum(v2_t, 2);

  % Return results
  y = [I1_RMS, I2_RMS, Pac_1, Pac_2, Pac_1 + Pac_2];

  % Time in microseconds for plots
  t_us = t * 1e6;

  % Plot voltage waveforms
  figure(1);
  plot(t_us, v1_total_t, 'b', 'DisplayName', '$v_{1}$', 'LineWidth', 1.5);
  hold on;
  plot(t_us, v2_total_t, 'r', 'DisplayName', '$v_{2}$', 'LineWidth', 1.5);
  xlabel('Time ($\mu$s)', 'Interpreter', 'latex', 'FontSize', 20);
  ylabel('Voltage (V)', 'Interpreter', 'latex', 'FontSize', 20);
  title('Voltage $v_{1}$ and $v_{2}$ with respect to time', 'Interpreter', 'latex',
'FontSize', 20);
  legend('Interpreter', 'latex', 'FontSize', 20);
  grid on;
  set(gca, 'FontSize', 20, 'TickLabelInterpreter', 'latex');
  xlim([0 10]);
  hold off;

  % Plot current waveforms
  figure(2);
  plot(t_us, i1_total_t, 'b', 'DisplayName', '$i_{1}$', 'LineWidth', 1.5);
  hold on;
  plot(t_us, i2_total_t, 'r', 'DisplayName', '$i_{2}$', 'LineWidth', 1.5);
  xlabel('Time ($\mu$s)', 'Interpreter', 'latex', 'FontSize', 20);
  ylabel('Current (A)', 'Interpreter', 'latex', 'FontSize', 20);
  title('Current $i_{1}$ and $i_{2}$ with respect to time', 'Interpreter', 'latex',
'FontSize', 20);
  legend('Interpreter', 'latex', 'FontSize', 20);
  grid on;
```

```
set(gca, 'FontSize', 20, 'TickLabelInterpreter', 'latex');
xlim([0 10]);
hold off;

% Display results
disp(['I1_RMS = ', num2str(I1_RMS), ' A']);
disp(['I2_RMS = ', num2str(I2_RMS), ' A']);
disp(['Pac_1 = ', num2str(Pac_1), ' W']);
disp(['Pac_2 = ', num2str(Pac_2), ' W']);
disp(['P_cond_loss = ', num2str(Pac_1 + Pac_2), ' W']);
```